Transient Electro-Thermal Modeling of Bipolar Power Semiconductor Devices

Synthesis Lectures on Power Electronics

Editor
Jerry Hudgins, *University of Nebraska, Lincoln*

Synthesis Lectures on Power Electronics will publish 50- to 100-page publications on topics related to power electronics, ancillary components, packaging and integration, electric machines and their drive systems, as well as related subjects such as EMI and power quality. Each lecture develops a particular topic with the requisite introductory material and progresses to more advanced subject matter such that a comprehensive body of knowledge is encompassed. Simulation and modeling techniques and examples are included where applicable. The authors selected to write the lectures are leading experts on each subject who have extensive backgrounds in the theory, design, and implementation of power electronics, and electric machines and drives.

The series is designed to meet the demands of modern engineers, technologists, and engineering managers who face the increased electrification and proliferation of power processing systems into all aspects of electrical engineering applications and must learn to design, incorporate, or maintain these systems.

Transient Electro-Thermal Modeling of Bipolar Power Semiconductor Devices
Tanya Kirilova Gachovska, Bin Du, Jerry L. Hudgins, and Enrico Santi
2013

Modeling Bipolar Power Semiconductor Devices
Tanya K. Gachovska, Jerry L. Hudgins, Enrico Santi, Angus Bryant, and Patrick R. Palmer
2013

Signal Processing for Solar Array Monitoring, Fault Detection, and Optimization
Mahesh Banavar, Henry Braun, Santoshi Tejasri Buddha, Venkatachalam Krishnan, Andreas Spanias, Shinichi Takada, Toru Takehara, Cihan Tepedelenlioglu, and Ted Yeider
2012

The Smart Grid: Adapting the Power System to New Challenges
Math H.J. Bollen
2011

Digital Control in Power Electronics
Simone Buso and Paolo Mattavelli
2006

Power Electronics for Modern Wind Turbines
Frede Blaabjerg and Zhe Chen
2006

Transient Electro-Thermal Modeling of Bipolar Power Semiconductor Devices

Tanya Kirilova Gachovska, Bin Du, Jerry L. Hudgins, and Enrico Santi

ISBN: 978-3-031-01378-2   paperback
ISBN: 978-3-031-02506-8   ebook

DOI 10.1007/978-3-031-02506-8

A Publication in the Springer series
SYNTHESIS LECTURES ON POWER ELECTRONICS

Lecture #6
Series Editor: Jerry Hudgins, *University of Nebraska, Lincoln*
Series ISSN
Synthesis Lectures on Power Electronics
Print 1931-9525   Electronic 1931-9533

Transient Electro-Thermal Modeling of Bipolar Power Semiconductor Devices

Tanya Kirilova Gachovska
Solantro Semiconductor Inc.

Bin Du
Danfoss Power Electronics

Jerry L. Hudgins
University of Nebraska

Enrico Santi
University of South Carolina

SYNTHESIS LECTURES ON POWER ELECTRONICS #6

ABSTRACT

This book presents physics-based electro-thermal models of bipolar power semiconductor devices including their packages, and describes their implementation in MATLAB and Simulink. It is a continuation of our first book *Modeling of Bipolar Power Semiconductor Devices*. The device electrical models are developed by subdividing the devices into different regions and the operations in each region, along with the interactions at the interfaces, are analyzed using the basic semiconductor physics equations that govern device behavior. The Fourier series solution is used to solve the ambipolar diffusion equation in the lightly doped drift region of the devices. In addition to the external electrical characteristics, internal physical and electrical information, such as junction voltages and carrier distribution in different regions of the device, can be obtained using the models.

The instantaneous dissipated power, calculated using the electrical device models, serves as input to the thermal model (RC network with constant and nonconstant thermal resistance and thermal heat capacity, or Fourier thermal model) of the entire module or package, which computes the junction temperature of the device. Once an updated junction temperature is calculated, the temperature-dependent semiconductor material parameters are re-calculated and used with the device electrical model in the next time-step of the simulation.

The physics-based electro-thermal models can be used for optimizing device and package design and also for validating extracted parameters of the devices. The thermal model can be used alone for monitoring the junction temperature of a power semiconductor device, and the resulting simulation results used as an indicator of the health and reliability of the semiconductor power device.

KEYWORDS

power semiconductor devices, physics-based model, Fourier series solution, drift region, carrier diffusion, transient switching behavior, thermal model, packaging, modules

*We dedicate this book to our families,
particularly Sheryl, Carol, Anna, and An,
who support our endeavors.
We would also like to thank Dr. Angus Bryant for
many helpful and stimulating conversations on the
topic of power device modeling.*

Contents

Nomenclature xi

1 Temperature Dependencies of Material and Device Parameters 1

  1.1 Introduction 1

  1.2 Temperature Dependencies 2

    1.2.1 Intrinsic Carrier Concentration 2

    1.2.2 Ionized Donor Impurity Concentration 3

    1.2.3 Carrier Mobility 4

    1.2.4 Lifetime 4

    1.2.5 Emitter Recombination Parameters 5

    1.2.6 Threshold Voltage and Transconductance 5

2 One-Dimensional Thermal Model 7

  2.1 Package Design 7

  2.2 Heat Conduction Problem in DBC structure 7

  2.3 Equivalent *RC* Network Thermal Model 10

  2.4 One-Dimensional Fourier Series Thermal Model 14

3 Realization of Power IGBT and Diode Thermal Model 23

  3.1 Introduction 23

  3.2 Realization of Equivalent *RC* Network 25

  3.3 Realization of One-Dimensional Fourier-Series Thermal Model 30

  3.4 Temperature Dependent Parameters of Diodes and Their Connection to an Electrical Model 34

  3.5 Temperature-Dependent Parameters of NPT IGBT and Their Connection to the Electrical Model 46

A Appendix 59

x

References 65

Authors' Biographies 67

Nomenclature

A	Active device area, cm^2
a	Current fall slope , As^{-1}
A_i, b_i, and m_i	Amplitude parameter (cm^{-1}), ionization energy parameter (Vcm^{-1}), and fit parameter
a_i	Ratio of intercell area to the total die area
b	Ratio of electron and hole mobilities
V_{BD}	Breakdown voltage, kV
C_{CG}, C_{GE}, and C_{dep}	Miller and gate emitter capacitance, μF
C_{CE} and C_{dep}	Collector emitter and depletion capacitance, μF
C_n and C_p	Auger coefficients, cm^6/s
C_J	Junction capacitance, μF
C_{J1} and C_{J2}	Junction capacitance at junction J_1 and J_2, μF
C_{oes}	Output capacitance provided in an IGBT data sheets, μF
C_{OX}	Oxide capacitance, μF
C_{ox}	Oxide capacitance per unit area, μFcm^2
C_{res}	Reverse transfer capacitance of a IGBT, μF
C_{sn}	Snubber capacitance, μF
d_1, W, and d_2	Depletion region thicknesses of each section of $N^+N^-P^+$ regions, respectively, cm
D	Ambipolar diffusivity, cm^{2}s^{-1}
D_{diff}	Diffusivity constant
D_n and D_p	Electron and hole diffusivities, cm^{2}s^{-1}
D_{n_P} and $D_{p_{N+}}$	Minority carrier diffusivities in P and N^+ region, cm^{2}s^{-1}
D_{pN}	Minority carrier diffusivities in N buffer layer, cm^{2}s^{-1}

D_{N^-}	Electron diffusivity in N^- region, cm^2s^{-1}
E_G	Band gap energy, eV
G	Generation rate, cm^3/s
$G(s)$	Transfer function
I_A i_{sn}, and i_L,	Diode, snubber, and load currents, A
I_F	DC forward current, A
h_n and h_p	Recombination parameters, $cm^{-4}s^{-1}$
I_B, I_C, I_E, and IG	Base, collector, emitter, and gate currents, A
i_C	Collector current, A
I_{CG}	Current due to Miller capacitance, A
I_{even} and I_{odd}	Currents for even and odd numbers of harmonic k, A
I_D	Diode currents, A
i_g	Gate currents, A
I_{mos}	Channel MOSFET current, A
I_{QN}	Capacitive current due to the variation in the charge stored in the buffer region, A
I_n and I_p	Electron and hole currents, A
I_{n0}, I_{n1}, and I_{n2}	Electron current at junctions J_0, J_1, and J_2, A
I_{p0}, I_{p1}, and I_{p2}	Electron current at junctions J_0, J_1, and J_2, A
I_{disp}, I_{disp1} and I_{disp2}	Displacement currents, A
i_{sn}	Snubber currents, A
J	Maximum current density, Acm^{-2}
J_0, J_1, and J_2,	Junctions 0, 1, and 2
J, J_p, and J_n	Total, hole, and electron current densities, Acm^{-2}
$J_{n|drift}$ and $J_{n|diff}$	Electron drift and diffusion current densities, Acm^{-2}
$J_{p|drift}$ and $J_{p|diff}$	Hole drift and diffusion current densities, Acm^{-2}
$J_{n(PN^-)}$ and $J_{p(N^+N^-)}$	Electron carrier densities at the PN^- and N^+N^- junctions, Acm^{-2}

k and n	Harmonic indexes for CSR carrier density of Fourier series representation
k	Boltzmann's constant (1.381×10^{-23} J/K)
K	Coefficient
K_{FV}	Feedback constant
K_p	Transconductance parameter, AV^{-2}
L_{ch} and W_{ch}	Channel length (cm) and width (cm) of the MOS structure
l_m and l_1	Intercell half-width and the width due to voltage V_{GE}, cm
L_{pN}	Diffusion length in N buffer layer, $cm^2 s^{-1}$
L_S	Stray inductance, H
M	Number of the terms of the Fourier series
n_{B0} and n_{B1}	Electron doping concentration in the P-base at the junction J_0 and J_1, cm^{-3}
N_{D2}, N_{D1}, N_A	Doping concentration in the $N^+ N^- P^+$ regions, respectively, cm^{-3}
N_{N^-}	Doping concentration in the N^- region, cm^{-3}
N_{SC}	Space charge density, cm^{-3}
n	Electron carrier concentration, cm^{-3}
n_i	Intrinsic carrier concentration, cm^{-3}
$n_0 \dots n_7$	Nodes in lumped-charge model
N	Buffer layer
N^- and N^+	Lightly doped and heavy doped hole regions
$N_{N+}, N_P,$ and N_N	Doping concentrations in N^+ and P region, and the buffer layer cm^{-3}
N_{N-} and N_H	Doping concentrations of the N^- drift region and buffer layer, cm^{-3}
N_A	Acceptor doping concentration, cm^{-3}
N_D	Donor doping concentration, cm^{-3}
p	Hole carrier concentration, cm^{-3}

P and P^+	Base and heavy doped hole regions
P_B	Doping concentration in base region, cm^{-3}
p_{N0} and p_{NW}	Hole concentrations at the two ends of the buffer layer, cm^{-3}
p_{N^--0}	Concentration in the beginning of the N^- drift region, cm^{-3}
$p(x)$	Ambipolar carrier density as a function of position, cm^{-3}
$p(x; t)$	Ambipolar carrier density as a function of position and time, cm^{-3}
$p_0(t)$	DC Fourier series component of the carrier storage region (CSR) carrier density profile, cm^{-3}
$p_k(t), p_n(t)$	k^{-th} - or n^{-th} Fourier series component of CSR carrier density profile, cm^{-3}
$p_{T1}, p_{T2},$ and p_T	Carrier densities of the two points of a segment and the carrier density of the segment, cm^{-3}
p_{T_k}	Carrier densities of the kth segment, cm^{-3}
$\tilde{p}_x$ and $\tilde{p}_k$	Carrier density profile in space and time domain
p_{x1} and p_{x2}	Excess carrier concentrations at the two boundaries of the drift region, cm^{-3}
q	Unit electrical charge ($\simeq 1.6 \times 10^{-19} \mathrm{C}$)
Q_B	Total electron charge in the base, C
Q_N	Charge stored in the buffer region, C
$Q_{RR}, I_{RM},$ and T_{RR}	Reverse recovery charge (C), reverse recovery peak current (A), and reverse recovery time (s)
R	Recombination rate, cm^3/s
R_g and L_g	Resistance (Ω) and inductance (H) in the gate circuit
R_L and L_L	Load resistance (Ω) and inductance (H)
R_{SRH}	Shockley-Read-Hall carrier recombination rate, $\mathrm{cm}^{-3}\mathrm{s}^{-1}$
R_{sn}	Snubber resistance, Ω
S	Softness factor
T	Absolute temperature, K

T_0	Time at which the diode current become equal to zero, s
T_1	Time at which the peak reverse current I_{RM} occurs, s
T_A	Difference between T_1 and T_0, s
V_{AK}	Voltage drop across the power diode, V
V_{appl}	Voltage applied to $P^+N^-N^+$ structure, V
V_{BR}	Breakdown voltage, V
V_{BE}	Base emitter voltage, V
V_{CE}	Collector emitter voltage, V
V_{DC}	DC voltage supplied to the circuit, V
V_d	Depletion layer voltage, V
V_{d1} and V_{d2}	Depletion layer voltages, V
V_{GE}	Gate emitter voltage, V
V_{GS}	Gate source voltage, V
V_{L_s}	Stray inductance (L_s) voltage, V
V_{j0}, V_{j1}, and V_{j2}	Voltage drop at junctions J_0, J_1, and J_2, V
v_{sat}	Saturation velocity, cm s^{-1}
V_{seg}	Voltage drop of a segment, V
V_T	Thermal voltage, V
V_{TH}	Threshold voltage, V
W_B	Base width, cm
W_d	Widths of the depletion region, cm
W_1 and W_2	Depletion widths, cm
W_{N-} and W_H	Widths of the depletion region on the N^- and buffer layer, cm
W_{d1} and W_{d2}	Depletion widths, cm
x_1 and x_2	Boundary positions of the carrier storage region, cm
α_i, and α_{eff}	General ionization coefficient, effective ionization coefficient, cm^{-1}

Δx	Segment thickness, cm
$\mathcal{E}$	Electric field strength, V/cm
ϵ	Material permittivity, Fcm^{-1}
μ_n and μ_p	Electron and hole mobilities, $(\text{cm}^2\text{V}^{-1}\text{s}^{-1})$
τ_B	High-level lifetime in the P- base region, s
τ_H	Carrier lifetime in N buffer-layer, s
τ_{lim}	Time constant, s
τ_n and τ_p	Electron and hole lifetimes, s
τ_{HL} and τ	High-level injection lifetimes, s
τ_{N^-}	Electron lifetime in N^- region, s
τ_{n_P} and $\tau_{p_{N^+}}$	Electron and hole lifetimes in P and N^+ region, s
τ_{pN}	Minority carrier lifetime in N buffer layer, cm^2s^{-1}
τ_{rr}	Reverse recovery time constant, s
AC	Alternating current
ADE	Ambipolar diffusion equation
BJT	Bipolar junction transistor
CSR	Carrier storage region
DC	Direct current
IGBT	Insulated gate bipolar transistor
MOS	Metal-oxide-semiconductor
MOSFET	Metal-oxide-semiconductor field-effect transistor
PT	Punch-through
Si	Silicon
SRH	Shockley-Read-Hall

C H A P T E R   1

Temperature Dependencies of Material and Device Parameters

1.1   INTRODUCTION

A simulation model is usually one of the principal tools used for designing and manufacturing power devices. A complete simulation model of power semiconductor devices should include thermal characterization and junction temperature prediction. The need for high-voltage, high-power-density devices operating at high frequency is growing, especially for advanced power electronics. As improved power semiconductor devices exhibit increased current density during the "on state" and increased blocking voltage capability during the "off state," these higher powered devices will necessarily have increased internal power loss, and an associated temperature rise due to the self-heating effect. Most semiconductor material parameters are temperature sensitive, and exposure to temperatures outside the recommended range will alter the operation of the power semiconductor device and can cause failure. It has been reported that nearly 60% of failures are temperature induced, and for every 10°C temperature rise in the operating environment, the failure rate nearly doubles. Therefore, junction temperature is an important factor in simulation models as a simulation based on a fixed temperature cannot properly characterize the real working conditions. Therefore, an electro-thermal model of power semiconductor devices is needed to accurately model the behavior of power semiconductor devices in a switching power converter application.

Thermal management is of great concern in terms of device performance, energy efficiency, and economic cost. The traditional design of a power electronics package uses an experimental thermal cycling test because of the difficulties in performing accurate transient power cycling simulation studies. Finding an exact thermal solution under the transient process is difficult. The finite element method (FEM) has been used to find such a solution, but while FEM is highly precise, the computing speed is slow. This solution method has been widely used for static and transient thermal analysis, and various commercial software packages that perform thermal analysis are based on it. Another major drawback of FEM is that it does not work well with certain electrical circuit simulators that could calculate the instantaneous transient power dissipation of power electronics components.

Another method of solving transient temperature problems is to build a thermal network based on the thermal resistance and capacitance of different packaging materials. This method has been adopted in most of the electrical models for power semiconductor devices. Using this method, the instantaneous junction temperature can be estimated with the help of circuit simulators, such as Spice and Saber. However, the temperature distribution in different package layers is unknown.

A power device model based on semiconductor physics has shown that it may provide an accurate solution to the dynamically varying carrier profile inside the device during transients. The carrier dynamics within the device during conduction under high-level injection are described by the ambipolar diffusion equation (ADE). The fundamental solution method used in this work is to calculate a Fourier series solution for the ADE. Because of the physical similarity between carrier diffusion and diffusive heat flow, the Fourier expansion is also used to solve the heat diffusion equation to determine heat conduction in different packaging materials. In this book, a general methodology and model based on the Fourier-series solution of the appropriate diffusion equation is described. As an example, the electro-thermal model of an IGBT under inductive load switching conditions is introduced. The model has been implemented in Matlab and Simulink.

1.2   TEMPERATURE DEPENDENCIES

The behavior of conductivity-modulated devices, such as diodes, thyristors, and IGBTs, depends significantly on the excess carrier (charge) distribution in the lightly doped drift region. The semiconductor material and device properties, such as intrinsic carrier concentration, electron and hole mobilities, carrier lifetime, MOS threshold voltage, and transconductance, are affected by temperature; therefore their temperature dependencies should be included in an electro-thermal device model. In the next section, relevant temperature dependencies of material properties and of IGBT parameters taken from a low-temperature study by Caiafa et al. in 2003 [1] are given. All parameters are for silicon devices.

1.2.1   INTRINSIC CARRIER CONCENTRATION

In a semiconductor in thermal equilibrium that is completely free of impurities and defects the electron and hole concentrations are equal. The intrinsic carrier concentration n_i is the number of electrons in the conduction band and of holes in the valence band per unit volume. The intrinsic carrier concentration can be expressed as (1.1) and is a function of the temperature, the electron and hole density-of-states effective-masses, m_n^* and m_p^*, and the energy bandgap, E_G. Note that m_n^*, m_p^*, and E_G are also temperature dependent.

$$n_i = 4.81 \times 10^{15} \left(\frac{m_n^* m_p^*}{m_0^2} \right)^{0.75} T^{1.5} e^{-\frac{E_G}{2kT}} . \tag{1.1}$$

In (1.1) m_0 is electron rest mass of Si and k is the Boltzmann constant.

Barber (1967) lists three mechanisms that are primarily responsible for effective mass temperature variation: (1) the temperature affects the lattice spacing, (2) the Fermi distribution depends on the temperature, and (3) temperature cause a variation in the energy band structure due to an interaction between the electrons and lattice vibrations. The electron and hole density-of-states effective-mass expressions for Si are given by Caiafa et al. (2003) [1]. They are presented as polynomial regression fits using the Barber data (1967). The expressions are valid for temperature values from 0–500 K:

$$m_n^* = (-1.084 \times 10^{-9}T^3 + 7.580 \times 10^{-7}T^2 + 2.862 \times 10^{-4}\text{T} + 1.057)m_0 \tag{1.2}$$

$$\begin{aligned} m_p^* = (1.872 \times 10^{-11}T^4 - 1.969 \times 10^{-8}T^3 + 5.857 \times 10^{-6}T^2 \\ + 2.712 \times 10^{-4}T + 0.584)m_0 \,. \end{aligned} \tag{1.3}$$

The energy bandgap of a semiconductor material, E_G, tends to decrease as the temperature in the material increases. The reason is that, as the temperature increases, the atomic vibration increases; and, therefore, the interatomic spacing also increases, which decreases the potential needed for an electron to jump from the valence to the conduction band. Bludau (1974) gives temperature dependencies of the bandgap, E_G, for the temperature range from 0–500 K:

$$\begin{aligned} E_G &= 1.17 + 1.059 \times 10^{-6}T - 6.05 \times 10^{-6}T^2 &\quad \text{for } T \leq 170 \text{ K} \\ E_G &= 1.1785 - 9.025 \times 10^{-5}T - 3.05 \times 10^{-7}T^2 &\quad \text{for } T > 170 \text{ K}\,. \end{aligned} \tag{1.4}$$

1.2.2   IONIZED DONOR IMPURITY CONCENTRATION

For more precise calculation of the voltage drop of the lightly doped drift region of a power semiconductor device during switching, it is necessary to include its ionized impurity doping concentration N_{N-}^+. But the impurity doping concentration is also temperature dependent, and it can be calculated by the following equation:

$$N_{N-}^+ = \frac{N_C}{2g_D} e^{\frac{E_G}{2kT}} \left[\left(1 + \frac{4N_{N-}}{N_C/g_D e^{\frac{E_G}{2kT}}} \right)^{\frac{1}{2}} - 1 \right] \tag{1.5}$$

where g_D is a donor degeneracy factor accounting for electron spin, usually assumed to be equal to two. The effective density of states, N_C, is calculated as:

$$N_C = n_i \left(\frac{m_n^*}{m_p^*} \right)^{0.75} e^{\frac{E_G}{2kT}} \,. \tag{1.6}$$

The charge due to the ionized doping concentration, N_{N-}^+, is:

$$Q_N = q N_{N-}^+ \tag{1.7}$$

where $q = 1.6 \times 10^{-19} C$ is electron charge.

1.2.3 CARRIER MOBILITY

Another important parameter for electrical performance of power switches is carrier mobility, which characterizes electron and hole transport due to the existing electric field. There are four scattering mechanisms that affect carrier mobility: lattice, donor, acceptor, and electron-hole scattering. The electron-electron and hole-hole scattering are always neglected due to them being second-order effects. An increase in semiconductor material temperature leads to an increase in scattering and number of collisions, therefore, the hole and electron mobility decrease. The temperature dependency of mobility is important for an accurate calculation of the voltage drop across the lightly doped drift region during the on-state and, therefore, to determine the on-state losses of a power semiconductor device. A good simulation model should include the temperature dependency of carrier mobility. The temperature dependencies of carrier electron and hole mobilities given by Caiafa *et al.* (2003) [1] are given below:

$$\mu_n = 2.92 \times 10^3 \left(\frac{T}{300}\right)^{-1.21}$$
$$\mu_p = 603 \left(\frac{T}{300}\right)^{-1.94}. \tag{1.8}$$

Einstein relationships for electrons and holes are:

$$\frac{D_n}{\mu_n} = \frac{kT}{q}$$
$$\frac{D_p}{\mu_p} = \frac{kT}{q} \tag{1.9}$$

where D_n and D_p are electron and hole diffusivities, respectively. From (1.8) and (1.9) it follows that D_n and D_p are also temperature dependent; therefore, their dependence should be included in a thermal model as well. Ambipolar diffusivity, D, is also temperature dependent, because it depends on the electron and hole diffusivities:

$$D = \frac{2D_n D_p}{D_n + D_p}. \tag{1.10}$$

Also, the diffusivity term, D_{diff}, given by Eq. (1.11) depends on electron and hole diffusivities; and, therefore, it is also temperature dependent:

$$D_{\text{diff}} = \frac{D_n - D_p}{D_n + D_p}. \tag{1.11}$$

1.2.4 LIFETIME

Carrier lifetime is a critical parameter in determining the dynamic behavior of power semiconductor switches. Usually for power switches in normal operating conditions, the low-doped base region is always under a high-level injection condition when the device is conducting. The associated effective carrier lifetime is approximately equal to the sum of the injected electron and hole minority carrier lifetimes, τ_n and τ_p. The temperature dependency of the carrier lifetime in the low-doped based region (for $n = p < 10^{17}$ cm^3) given by Caiafa et al. (2003) [1] is:

$$\tau = \tau_0 \left(\frac{T}{300}\right)^{0.57} \left[1 + \frac{\left(\frac{T}{300}\right)^{1.2} - 1}{0.6276 + 149\tau_0 + \sqrt{2.22 \times 10^4 (\tau_0^2 - 5 \times 10^{-3}\tau_0) + 0.3938}} \right] \quad (1.12)$$

where τ_0 is the lifetime at a temperature of 300 K.

1.2.5   EMITTER RECOMBINATION PARAMETERS

Emitter recombination parameters h_n and h_p are a function of the emitter's doping concentrations, the minority carrier diffusivities, and lifetime; therefore, they are also dependent on temperature. The temperature dependencies are:

$$h_n = h_{n_0} \left(\frac{300}{T_0}\right)^{2.5} \quad (1.13)$$

$$h_p = h_{p_0} \left(\frac{300}{T_0}\right)^{2.5} \quad (1.14)$$

where h_{n_0} and h_{p_0} are the emitter recombination parameters at a temperature of 300 K.

1.2.6   THRESHOLD VOLTAGE AND TRANSCONDUCTANCE

For power devices with a metal-oxide-semiconductor field-effect transistor (MOSFET) gate structure, such as power MOSFETs or IGBTs, the threshold voltage and transconductance are very important parameters. Caiafa et al. (2003) [1] give the temperature dependence of the threshold voltage as:

$$V_{Th} = V_{Th_0} - 1.13 \times 10^{-3}(T - 300) \quad (1.15)$$

where V_{Th_0} is the threshold voltage at a temperature of 300 K.

Caiafa et al. [1] also give a temperature dependence for the transconductance parameter of an IGBT as:

$$K_p = K_{p0} \left(\frac{T}{300}\right)^{-0.95} \quad (1.16)$$

where K_{p0} is the transconductance parameter at a temperature of 300 K.

C H A P T E R   2

One-Dimensional Thermal Model

2.1   PACKAGE DESIGN

Packaging technology plays an important role in power electronics manufacturing. Power semiconductor switches usually need a good thermally conducting package to transfer the dissipated heat to the heat sink efficiently and keep the junction temperature under the maximum permissible value. A good package should handle high power dissipation and provide long life and high reliability. The package also provides solid mechanical support for the die and galvanic isolation between the die and heat sink for safety reasons. The accurate thermal analysis of a power semiconductor device necessarily depends on the thermal behavior of the package structure.

Power semiconductor devices, such as MOSFETs and IGBTs, are traditionally packaged in the structure shown in Fig. 2.1. The silicon die is soldered to a direct-bonded copper (DBC) substrate. The DBC substrate contains dielectric material, AlN, sandwiched between two layers of copper. The AlN provides electrical isolation between the silicon die and the external device case. DBC is soldered to a base plate, usually made of copper. To increase the heat exchange area, the base plate is then bolted onto a large heat sink having a large fin structure. Between the base plate and the heat sink, a thin layer of thermal grease is added to reduce contact thermal resistance. Heat is generated in the die and then transferred through the solder, copper, dielectric, another copper layer, an additional solder layer, to the copper baseplate, and finally out through the heat sink to the ambient environment. The total thermal resistance is the combination of the thermal resistances of all the material layers in the thermal conducting path. The package design is such that under proper operation of the power device, its junction temperature is maintained below the maximum permissible value.

2.2   HEAT CONDUCTION PROBLEM IN DBC STRUCTURE

The heat generated from the heat source, the semiconductor die, is transferred to the heatsink to be dissipated into the environment, as shown in Fig. 2.2(a). Considering a symmetrical structure, only half of the package needs to be simulated, as illustrated in Fig. 2.2(b). The heat transferred from an electronic component die into the heat sink is dominated by conduction, so that convection and radiation are assumed to be negligible. For a homogeneous, isotropic layer, the heat

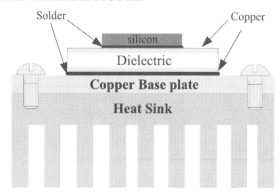

Figure 2.1: The structure of a DBC package.

transfer is expressed by the following heat conduction equation:

$$\nabla(k_i \cdot \nabla T_i(x,t)) + g_i = \rho_i c_i \frac{\partial T_i(x,t)}{\partial t} \qquad \text{for } i = 1 \ldots m \qquad (2.1)$$

where i is the material layer number, k_i is the thermal conductivity in the direction of heat conduction, ρ_i and c_i are the mass density and specific heat of the ith layer, respectively, and g_i is the heat generation function within the ith layer. If heat input is only assumed to be input at the top boundary of the silicon chip, g_i is zero (only external heat source). Assuming that the surrounding temperature T_0 is a constant, the temperature $T(x,t)$ can be replaced by the temperature difference $T(x,t) - T_0$.

Since the width of each layer is much larger than its thickness, it is often sufficient to model the structure as one-dimensional (1-D). The simplified 1-D model is illustrated in Fig. 2.3.

If the thermal coefficients, the thermal conductivity and heat capacity of each layer are considered as constants, the governing equations are simplified as:

$$\alpha_i \frac{\partial^2 T_i(x,t)}{\partial x^2} + \frac{g_i(x,t)}{\rho_i c_i} = \frac{\partial T_i(x,t)}{\partial t} \qquad \text{for } i = 1 \ldots m \qquad (2.2)$$

where α_i is the thermal diffusivity of the ith layer, which is given by $\alpha_i = k_i / (\rho_i c_i)$. Assuming a perfect thermal contact between the different layers, the interface boundary conditions require the temperatures to be continuous:

$$T_i(x_i,t) = T_{i+1}(x_i,t) \qquad \text{for } i = 1 \ldots m \qquad (2.3)$$

and the heat fluxes passing through different layers must be continuous:

$$k_i \frac{\partial T_i(x,t)}{\partial x}\bigg|_{x_i} = k_{i+1} \frac{\partial T_{i+1}(x,t)}{\partial x}\bigg|_{x_i} \qquad \text{for } i = 1 \ldots m \,. \qquad (2.4)$$

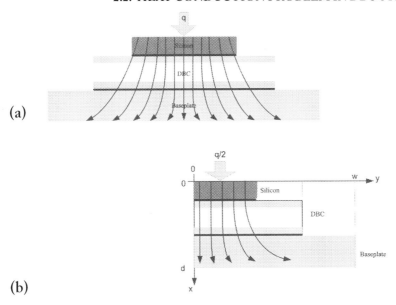

Figure 2.2: (a) Heat dissipation path and (b) the simulation domain in DBC structures.

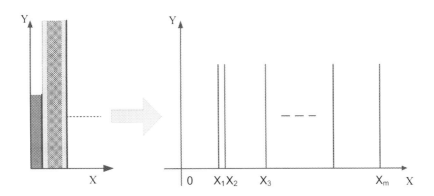

Figure 2.3: One-dimensional heat conduction in DBC structure.

For the initial condition, it is assumed that the whole package is at room temperature, which is 300 K. To solve Eq. (2.2) explicitly in the region $[x_1, x_m]$, boundary conditions at x_1 and x_m are necessary. If there is no heat generation inside the package, i.e., all of the heat is flowing from the bottom boundary of the silicon chip (coordinate x_1), the heat source is located only on the bottom side of the silicon layer, which is expressed as:

$$\left.\frac{\partial T_1(x,t)}{\partial x}\right|_{x=x_1} = \frac{q(t)}{k_1}. \tag{2.5}$$

If there are heat sources inside the package, the boundary condition Eq. (2.5) cannot be used without including the g_i term in (2.2).

For surfaces exposed to a fluid with a different temperature, convectional heat transfer happens between the surface and the fluid. This occurs at the surface of the heatsink, which corresponds to position x_m along the one-dimensional DBC structure. The convection surface boundary condition is expressed as:

$$k_m \left.\frac{\partial T_m(x,t)}{\partial x}\right|_{x=x_{m+1}} + h(T_m(x_m,t) - T_0) = 0 \tag{2.6}$$

where m represents the bottom layer number, and h is the convection heat transfer coefficient (W/(m²·K)). If $h = 0$, no heat flux passes through the bottom layer, which means there is a perfect thermal insulator at the bottom. If $h = \infty$, the temperature at the interface is equal to the ambient temperature, i.e., $T_m(x_m,t) = T_0$. Most of the time, a cooling system is used to achieve a near constant temperature at the outside case surface.

The convection heat transfer coefficient h is determined by assuming natural, free convection in air. At room temperature, the convection heat transfer coefficient, h, is 5–25 W/m²·K. Natural convection is always used to dissipate heat flows of 150–1500 W/cm². For higher heat dissipation, other cooling techniques, like forced air-cooling and liquid cooling, should be used.

2.3   EQUIVALENT *RC* NETWORK THERMAL MODEL

One easy method used to solve transient temperature problems is to build up an equivalent thermal network based on the thermal property of different packaging materials. The method is valid for a steady-state heat transfer in a medium with no heat generation and constant thermal conductivity. A schematic of a Foster-Equivalent RC thermal network is shown in Fig. 2.4. The equivalent current source $P(t)$ is equal to the instantaneous power dissipated in the die. Each layer is represented by the parallel combination of a thermal resistance R_i and a thermal capacitance C_i. The instantaneous voltage at each node is numerically equal to the temperature rise over ambient temperature, represented by the ground potential.

Usually, a semiconductor device manufacturer provides an experimental transient thermal response or transient thermal impedance curve, as shown in Fig. 2.5. The curve is a measure of the semiconductor temperature (junction temperature) and is referenced to the device's external case or ambient temperature. The x-axis represents the time duration of an input heat power pulse applied to the die and y-axis plots the maximum temperature rise per unit of applied power. This quantity is called the transient thermal impedance and is defined as:

$$Z_{thjc}(t) = \frac{T_{jMAX} - T_c}{P} = \frac{\Delta T_{jc}}{P} \tag{2.7}$$

where P is the pulse amplitude, t is the pulse duration T_{jMAX} is maximum junction temperature and T_c is the case temperature. As pulse duration t increases, the transient thermal impedance approaches asymptotically the thermal resistance from junction to case.

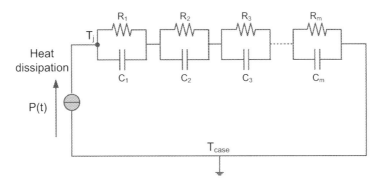

Figure 2.4: Foster-equivalent RC thermal network.

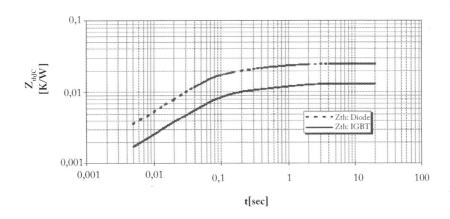

Figure 2.5: Experimental curves of transient thermal impedance for a diode and an IGBT.

To calculate *RC* parameters of a thermal network, a curve fitting method is applied to the experimental impedance data that results in a finite sum of exponential terms:

$$Z_{thjc} = \sum_{i=1}^{n} R_i \left(1 - e^{-\frac{t}{\tau_i}}\right), \tag{2.8}$$

where i is the term index, R_i and τ_i are correspondingly the thermal resistance and the thermal time constant of the ith term. The thermal time constant, τ_i can be used for calculating thermal capacitance, C_i, by following equation:

$$\tau_i = R_i C_i .$$

(2.9)

Generally, a semiconductor device manufacturer provides values for the thermal resistances and thermal time constants up to order 4 (e.g., i = 1, 2, 3, and 4). In this case, the four R-C cells do not necessarily correspond to different material layers. A disadvantage of the Foster-Equivalent RC thermal network is that the circuit shows only a behavioral characteristic of the system, and does not correlate directly with the physical parameters of the package materials and geometry. Furthermore, this topology does not correspond to physical reality where heat flow (and corresponding temperature rise) to the device case (copper layer) is delayed due to the inherent heat capacity and thermal resistance of various layers away from the die. The Cauer RC network accounts for the time delay and provides a better physical description/topology of the heat flow path. The Cauer network can be obtained from the Foster network so that the two networks are equivalent between the T_j and T_{case} nodes. However, the mathematical representation of the Cauer network is much more complicated as compared to the Foster network. The component equations to convert between the Foster and Cauer networks are provided in the Appendix.

In order to obtain the thermal resistance and capacitance directly from the geometry and materials used in a physical power semiconductor package, the Cauer equivalent RC network shown in Fig. 2.6 should be used. Each group of R and C represents different layers of the package between the heat source and the base plate. The heatsink could also be considered by simply adding two more RC component sections to represent the heat transfer from case-to-heatsink and from heatsink-to-ambient, respectively.

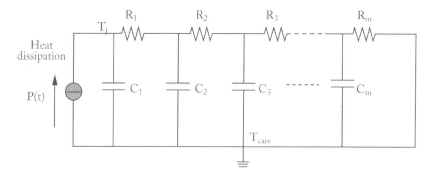

Figure 2.6: The Cauer equivalent RC thermal network.

The conduction thermal resistance, R_i, of the ith layer is calculated by:

$$R_i = \frac{d_i}{k_i A_i}$$

(2.10)

where d_i is the layer thickness, A_i is the effective cross-section area, and k_i is the thermal conductivity of the ith layer. For calculating the effective cross-section area, A_i, a first-order approximation can be used for the heat spreading through the layers based on a spreading angle of $45°$, centered on the heat source, in a conductor and no spreading angle ($0°$) in an insulator.

An equivalent resistance of the heatsink, R_m, can be obtained as the sum of a resistance due to the conductive heat transfer, calculated using Eq. (2.10), and a resistance due to the convective heat transfer. The latter represents a thermal exchange between the heat sink and the surrounding fluid (typically air); it is usually a nonlinear resistance, because both natural and forced convection are a nonlinear function of temperature difference. The heatsink resistance due to convective heat transfer, R_{m_C}, is inversely proportional to the convective heat transfer coefficient of the sink, h, and surface area exposed to the fluid, A:

$$R_{m_C} = \frac{1}{hA}. \tag{2.11}$$

The thermal capacitance, C_i, of the ith layer is calculated by the equation:

$$C_i = c_i \rho_i V_i \tag{2.12}$$

where c_i is the specific heat capacity, ρ_i is the mass density, and V_i is the effective volume of the ith layer.

The two coefficients, thermal conductivity, k_i, and specific heat capacity, c_i, are temperature dependent. The measured temperature variation of the thermal conductivity, k_i, and the specific heat capacity, c_i, of different materials has been used to create a least squares regression fit for the thermal conductivity (power curve) and the heat capacity (quadratic expression). The results for the different layers of the DBC package shown in Fig. 2.1 are presented in Table 2.1.

Table 2.1: Conductivity and specific heat capacity of the different material layers

Material	Thermal Conductivity (k_i) Power Fit: $k_i = aT^b$			Specific Heat Capacity (c_i) Quadratic Fit: $c_i = a + bT + dT^2$			
	a	b	Ref	a	b	d	Ref
Copper	608.0409	-0.07305	[2]	1.78E+02	2.36E+00	-2.06E-03	[2]
Silicon	438056.1	-1.40022	[2]	2.93E+02	3.96E-01	-3.20E-04	[2]
AlN	1.31E+08	-2.2626	[3]	819.7	0	0	[4]
Solder	241.8863	-0.22605	[4]	134	0	0	[5]

The input current source of the *RC* thermal network is the heat dissipation generated from a semiconductor chip. It is the quantity that links the thermal and electrical simulation models. The heat (power dissipated) term is calculated at each time step in the electrical simulation as the product of the device (semiconductor) voltage drop multiplied by its conducted current.

To calculate the junction temperature by using an RC equivalent thermal network, two similarities between thermal and electric RC networks can be used. The first is that the temperature difference across a layer generates heat transfer just as the potential difference across an electrical resistance causes current to flow. The second is that the conduction heat transfer rate in a layer is from high to low temperature, similar to the current flowing in a resistor from high to low potential. The junction temperature is modeled similar to a node voltage in an electrical system, so that the thermal model has the form of an R-C electrical network and it can be simulated with the same circuit software as the electrical model. The temperature dependency of thermal conductivity and heat capacity can be included in the thermal model, which will decrease the error during high dissipation conditions, but will make the model more complicated and lead to longer simulation times, and possibly convergence problems in the numerical simulation.

2.4   ONE-DIMENSIONAL FOURIER SERIES THERMAL MODEL

The one-dimensional (1-D) heat diffusion equation can be solved using a Fourier-series solution for the dynamic (time-dependent) temperature distribution, This Fourier-series solution can be represented as a 1-D thermal model that computed the transient thermal response. The approach taken by the model is to solve the thermal diffusion equation, describing the dynamic (time-dependent) temperature distribution (over the package geometry shown in Fig. 2.7 for $x>0$), with a Fourier series solution. The package is expanded for $x<0$ symmetrically about the vertical axis to simplify the Fourier series expansion.

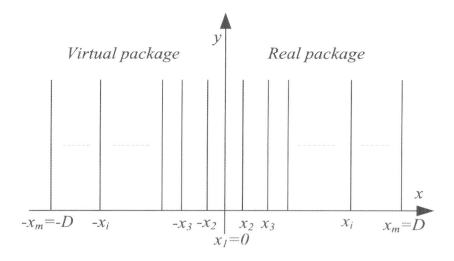

Figure 2.7: Simulation domain of 1-D heat conduction.

A conductive heat flow equation used to develop the transient thermal model was provided as a partial differential equation, Eq. (2.1). The 1-D form of this heat conduction equation is given by Eq. (2.2). It can be simplified from the general form of a heat conduction equation as the internal heat generation function, g, is equal to zero inside in the module layers. The Fourier expansion of the temperature distribution is given below:

$$T(x,t) = T_0(t) + \sum_{k=1}^{\infty} T_k(t) \cos k\omega_0 x \tag{2.13}$$

where the period of the fundamental harmonic of the Fourier series is equal to 2-D, and therefore the fundamental frequency is:

$$\omega_0 = \frac{2\pi}{2D} = \frac{\pi}{D} . \tag{2.14}$$

The heat diffusion equation in each material layer, i, is given:

$$\alpha_i \frac{\partial^2 T_i(x,t)}{\partial x^2} = \frac{\partial T_i(x,t)}{\partial t} . \tag{2.15}$$

The temperature $T_i(x,t)$ is the temperature distribution in the ith material layer. It is bolded in order to help the reader identify when the temperature function is both spatially and temporally dependent.

Using the same form as before for the Fourier series expansion, the heat-conduction equation can be solved by assuming that the solution of the temperature in the ith layer is:

$$T_i(x,t) = T_{i_0}(t) + \sum_{k=1}^{\infty} T_{i_k}(t) \cos k\omega_0 x \tag{2.16}$$

with ω_0 remaining as the fundamental frequency of the entire structure as expressed above in Eq. (2.14). Note that $T_i(x,t)$ is non-zero only in the interval $[x_i, x_{i+1}]$. Each harmonic amplitude, $T_{i_k}(t)$, is determined by the following equations:

$$T_{i_0}(t) = \frac{1}{D} \int_{x_i}^{x_{i+1}} T_i(x,t)\, dx$$

$$T_{i_k}(t) = \frac{2}{D} \int_{x_i}^{x_{i+1}} T_i(x,t) \cos \frac{k\pi x}{D}\, dx . \tag{2.17}$$

The Fourier series solution for the entire DBC package, obtained by summing the solutions of every DBC layer, is:

$$T(x,t) = \sum_{i=1}^{m} T_{i_0}(t) + \sum_{i=1}^{m} \sum_{k=1}^{\infty} T_{i_k}(t) \cos k\omega_0 x . \tag{2.18}$$

Substituting Eq. (2.17) into Eq. (2.18) and simplifying the Fourier series for the entire DBC, the following equations can be obtained:

$$T(x,t) = T_0(t) + \sum_{k=1}^{\infty} T_k(t) \cos k\omega_0 x \tag{2.19}$$

where the harmonic amplitudes, $T_0(t)$ and $T_k(t)$, are determined then by the following equations:

$$T_0(t) = \frac{1}{2D} \int_{-x_m}^{x_m} T(x,t)dx = \frac{1}{D} \int_{x_1}^{x_m} T(x,t)\,dx$$

$$= \frac{1}{D} \left[\int_{x_1}^{x_2} T(x,t)dx + \int_{x_2}^{x_3} T(x,t)dx + \ldots + \int_{x_1}^{x_m} T(x,t)\,dx \right]$$

$$= \frac{1}{D} \sum_{i=1}^{m} \int_{x_{i-1}}^{x_i} T(x_i,t)\,dx$$

$$= \sum_{i=1}^{m} T_{i_0}(t) \tag{2.20}$$

$$T_k(t) = \frac{2}{2D} \int_{-x_m}^{x_m} T(x,t) \cos \frac{k\pi x}{D}\,dx = \frac{2}{D} \int_{x_1}^{x_m} T(x,t) \cos \frac{k\pi x}{D}\,dx$$

$$= \frac{2}{D} \left[\int_{x_1}^{x_2} T(x,t) \cos \frac{k\pi x}{D}\,dx + \int_{x_2}^{x_3} T(x,t) \cos \frac{k\pi x}{D}\,dx \ldots + \int_{x_{m-1}}^{x_m} T(x,t) \cos \frac{k\pi x}{D}\,dx \right]$$

$$= \sum_{i=1}^{m} \frac{2}{D} \int_{x_{i-1}}^{x_i} T(x,t) \cos \frac{k\pi x}{D}\,dx$$

$$= \sum_{i=1}^{m} T_{i_k}(t) .$$

Again, the 1-D form of the heat conduction equation of the entire DBC is:

$$\alpha(x) \frac{\partial^2 T(x,t)}{\partial x^2} = \frac{\partial T(x,t)}{\partial t} \tag{2.21}$$

where $\alpha(x)$ is the thermal diffusivity, which is piecewise constant (it is constant within each material layer). To find the Fourier series solution for the entire DBC, each term of Eq. (2.21) is multiplied by $\cos\left(\frac{k\pi x}{D}\right)$ and then integrated with respect to x between x_1 and x_m:

$$\int_{x_1}^{x_m} \alpha \frac{\partial^2 T(x,t)}{\partial x^2} \cos\left(\frac{k\pi x}{D}\right) dx = \int_{x_1}^{x_m} \frac{\partial T(x,t)}{\partial t} \cos\left(\frac{k\pi x}{D}\right) dx . \tag{2.22}$$

The left-hand side of this equation is defined as I_1 and the right-hand side as I_2.

Due to the fixed geometry of the DBC structure, the right-hand side of Eq. (2.22) is a simplified form of the Fourier series transformation, and for

$$k \neq 0 :$$

$$I_2 = \int_{x_1}^{x_m} \frac{\partial T(x,t)}{\partial t} \cos\left(\frac{k\pi x}{D}\right) dx = \frac{D}{2} \frac{\partial T_k(t)}{\partial t}$$

$$\tag{2.23}$$

$$k = 0 :$$

$$I_2 = \int_{x_1}^{x_m} \frac{\partial T(x,t)}{\partial t} \cos\left(\frac{k\pi x}{D}\right) dx = D \frac{\partial T_k(t)}{\partial t} .$$

For the left-hand side of Eq. (2.22), the integral with respect to x is broken up into separate integrals for each layer (layer i corresponds to the integral between x_i and x_{i+1}) all added together. The expression in the second row is obtained integrating by parts twice:

$$I_1 = \sum_{i=1}^{m-1} \int_{x_i}^{x_{i+1}} \alpha_i \frac{\partial^2 T(x,t)}{\partial x^2} \cos\left(\frac{k\pi x}{D}\right) dx$$

$$= \sum_{i=1}^{m-1} \left\{ \alpha_i \cos\left(\frac{k\pi x}{D}\right) \frac{\partial T(x,t)}{\partial x} \bigg|_{x=x_i}^{x=x_{i+1}} + \frac{k\pi}{D}\alpha_i \sin\left(\frac{k\pi x}{D}\right) T(x,t) \bigg|_{x=x_i}^{x=x_{i+1}} \right. \tag{2.24}$$

$$\left. - \left(\frac{k\pi}{D}\right)^2 \int_{x_i}^{x_{i+1}} T(x,t)\alpha_i \cos\left(\frac{k\pi x}{D}\right) dx \right\}$$

$$= I_{11} + I_{12} + I_{13}$$

where k^{th} subscript refers to the index of the specific harmonic and i^{th} subscript refers to the material layer, and I_{11}, I_{12}, and I_{13} are expressed by the following equations:

$$I_{11} = \sum_{i=1}^{m-1} \alpha_i \cos\left(\frac{k\pi x}{D}\right) \frac{\partial T(x,t)}{\partial x} \Bigg|_{x=x_i}^{x=x_{i+1}} \tag{2.25}$$

$$I_{12} = \sum_{i=1}^{m-1} \frac{k\pi}{D} \alpha_i \sin\left(\frac{k\pi x}{D}\right) T(x,t) \Bigg|_{x=x_i}^{x=x_{i+1}} \tag{2.26}$$

$$I_{13} = -\left(\frac{k\pi}{D}\right)^2 \sum_{i=1}^{m-1} \int_{x_i}^{x_{i+1}} T(x,t)\, \alpha_i \cos\left(\frac{k\pi x}{D}\right) dx \,. \tag{2.27}$$

According to Eq. (2.13), the temperature partial differential at $x = x_i$ can be obtained by the following equations:

$$\frac{\partial T(x,t)}{\partial x}\Bigg|_{x=x_i} = -\frac{k\pi}{D} \sum_{i=1}^{\infty} T_k(t) \sin\left(\frac{k\pi x_i}{D}\right). \tag{2.28}$$

At the internal surface of two different layers, the interface boundary condition (2.4) (heat flux continuity) can be expressed as:

$$A_i k_i \frac{\partial T_i(x,t)}{\partial x}\Bigg|_{x=x_i^-} = A_{i+1} k_{i+1} \frac{\partial T_{i+1}(x,t)}{\partial x}\Bigg|_{x=x_i^+}$$

$$\frac{\partial T_{i+1}(x,t)}{\partial x}\Bigg|_{x=x_i^+} = C_{i,i+1} \frac{\partial T_i(x,t)}{\partial x}\Bigg|_{x=x_i^-} \tag{2.29}$$

where $C_{i,i+1} = (A_i k_i) / (A_{i+1} k_{i+1})$.

Note that the temperature gradient is different at the two sides of an interface. The temperature gradient at the interface left side can be written as Eq. (2.28) and the gradient of right side can be expressed by Eq. (2.29). Then, the integral I_{11}, Eq. (2.25), can be given as:

$$
\begin{aligned}
I_{11} &= \sum_{i=1}^{m-1} \alpha_i \cos\left(\frac{k\pi x}{D}\right) \frac{\partial T(x,t)}{\partial x}\bigg|_{x=x_i^+}^{x=x_{i+1}^-} \\
&= -\alpha_1 \frac{\partial T(x,t)}{\partial x}\bigg|_{x=x_1^+} + \alpha_1 \cos\left(\frac{k\pi x_2}{D}\right) \frac{\partial T(x,t)}{\partial x}\bigg|_{x=x_2^-} \\
&\quad + \alpha_2 \cos\left(\frac{k\pi x_3}{D}\right) \frac{\partial T(x,t)}{\partial x}\bigg|_{x=x_3^-} \\
&\quad - \alpha_2 \cos\left(\frac{k\pi x_2}{D}\right) \frac{\partial T(x,t)}{\partial x}\bigg|_{x=x_2^+} + \cdots + \alpha_{m-1}(-1)^k \frac{\partial T(x,t)}{\partial x}\bigg|_{x=x_m^-} \\
&\quad - \alpha_{m-1} \cos\left(\frac{k\pi x_{m-1}}{D}\right) \frac{\partial T(x,t)}{\partial x}\bigg|_{x=x_{m-1}^+} \\
&= \alpha_{m-1}(-1)^k \frac{\partial T(x,t)}{\partial x}\bigg|_{x=x_m^-} - \alpha_1 \frac{\partial T(x,t)}{\partial x}\bigg|_{x=x_1^+} \\
&\quad + \sum_{i=1}^{m-2} (\alpha_i - C_{i,i+1}\alpha_{i+1}) \cos\left(\frac{k\pi x_{i+1}}{D}\right) \frac{\partial T(x,t)}{\partial x}\bigg|_{x=x_{i+1}^-} \\
&= \alpha_{m-1}(-1)^k \frac{\partial T(x,t)}{\partial x}\bigg|_{x=x_m^-} - \alpha_1 \frac{\partial T(x,t)}{\partial x}\bigg|_{x=x_1^+} \\
&\quad - \sum_{i=1}^{m-2} (\alpha_i - C_{i,i+1}\alpha_{i+1}) \left\{ \sum_{n=1}^{\infty} \frac{n\pi}{D} \sin\left(\frac{n\pi x_{i+1}}{D}\right) \cos\left(\frac{k\pi x_{i+1}}{D}\right) T_n(t) \right\} \\
&= \alpha_{m-1}(-1)^k \frac{\partial T(x,t)}{\partial x}\bigg|_{x=x_m^-} - \alpha_1 \frac{\partial T(x,t)}{\partial x}\bigg|_{x=x_1^+} \\
&\quad - \sum_{i=1}^{m-2} \sum_{n=1}^{\infty} \frac{n\pi}{D} (\alpha_i - C_{i,i+1}\alpha_{i+1}) \sin\left(\frac{n\pi x_{i+1}}{D}\right) \cos\left(\frac{k\pi x_{i+1}}{D}\right) T_n(t)
\end{aligned}
\tag{2.30}
$$

where the kth and n^{th} subscripts refer to the index of the specific harmonic and the ith subscript refers to the material layer.

Assuming a perfect thermal contact at the interface, which means the temperature at the interface is the same for both sides, the integral I_{12} of Eq. (2.26) can be expanded as:

$$I_{12} = \frac{k\pi}{D} \sum_{i=1}^{m-1} \alpha_i \sin\left(\frac{k\pi(x)}{D}\right) T(x,t) \Big|_{x_i}^{x_{i+1}}$$

$$= \frac{k\pi}{D} \left\{ \begin{array}{l} 0 + \alpha_1 \sin\left(\frac{k\pi x_2}{D}\right) T(x_2,t) - \alpha_2 \sin\left(\frac{k\pi x_2}{D}\right) T(x_2,t) + \alpha_2 \sin\left(\frac{k\pi x_3}{D}\right) T(x_3,t) \\ -\alpha_3 \sin\left(\frac{k\pi x_3}{D}\right) T(x_3,t) + \cdots + 0 - \alpha_{m-1} \sin\left(\frac{k\pi x_{m-1}}{D}\right) T(x_{m-1},t) \end{array} \right\}$$

$$= \frac{k\pi}{D} \sum_{i=1}^{m-2} (\alpha_i - \alpha_{i+1}) \sin\left(\frac{k\pi x_{i+1}}{D}\right) T(x_{i+1},t) \qquad (2.31)$$

$$= \frac{k\pi}{D} \sum_{i=1}^{m-2} (\alpha_i - \alpha_{i+1}) \left\{ \sum_{n=0}^{\infty} \sin\left(\frac{k\pi x_{i+1}}{D}\right) \cos\left(\frac{n\pi x_{i+1}}{D}\right) T_n(t) \right\}$$

The integral I_{13}, Eq. (2.27), is complicated due to integration over different layers. After certain simplifications the integral I_{13} can be given in the following form:

$$I_{13} = -\left(\frac{k\pi}{D}\right)^2 \sum_{i=1}^{m-1} \int_{x_i}^{x_{i+1}} T(x,t)\alpha_i \cos\left(\frac{k\pi x}{D}\right) dx$$

$$= -\left(\frac{k\pi}{D}\right)^2 \sum_{i=1}^{m-1} \left\{ \int_{x_i}^{x_{i+1}} \sum_{n=0}^{\infty} \alpha_i \cos\left(\frac{n\pi x}{D}\right) \cos\left(\frac{k\pi x}{D}\right) T_n(t) dx \right\} \qquad (2.32)$$

$$= -\left(\frac{k\pi}{D}\right)^2$$

$$\sum_{i=1}^{m-1} \left\{ \begin{array}{l} \frac{\alpha_i}{2}\left[x_{i+1} - x_i + \frac{D}{2k\pi}\left(\sin\left(\frac{2k\pi x_{i+1}}{D}\right) - \sin\left(\frac{2k\pi x_i}{D}\right)\right)\right] T_k(t) \\ + \sum_{\substack{n=0 \\ n \neq k}}^{\infty} \frac{\alpha_i}{2}\left[\begin{array}{l} \frac{D}{(n+k)\pi}\left(\sin\left(\frac{(n+k)\pi x_{i+1}}{D}\right) - \sin\left(\frac{(n+k)\pi x_i}{D}\right)\right) \\ + \frac{D}{(n-k)\pi}\left(\sin\left(\frac{(n-k)\pi x_{i+1}}{D}\right) - \sin\left(\frac{(n-k)\pi x_i}{D}\right)\right) \end{array} \right] T_n(t) \end{array} \right\}$$

Combine Eq. (2.23), Eq. (2.30), Eq. (2.31), and Eq. (2.32) together; the post-transformation form Eq. (2.21) can be written as the following equation:

for $k > 0$

$$\frac{D}{2} \frac{\partial T_k(t)}{\partial t} = \alpha_{m-1}(-1)^k \left. \frac{\partial T(x,t)}{\partial x} \right|_{x=x_m^-} - \alpha_1 \left. \frac{\partial T(x,t)}{\partial x} \right|_{x=x_1^+}$$

$$-\frac{\pi}{D} \sum_{i=1}^{m-2}$$

$$\left\{ \sum_{n=1}^{\infty} \left[\begin{array}{l} n\,(\alpha_i - C_{i,i+1}\alpha_{i+1}) \sin\left(\frac{n\pi x_{i+1}}{D}\right) \cos\left(\frac{k\pi x_{i+1}}{D}\right) \\ -k\,(\alpha_i - \alpha_{i+1}) \sin\left(\frac{k\pi x_{i+1}}{D}\right) \cos\left(\frac{n\pi x_{i+1}}{D}\right) \end{array} \right] T_n(t) \right\}$$

$$-\left(\frac{k\pi}{D}\right)^2 \sum_{i=1}^{m-1}$$

$$\left\{ \begin{array}{l} \frac{\alpha_i}{2}\left[x_{i+1} - x_i + \frac{D}{2k\pi}\left(\sin\left(\frac{2k\pi x_{i+1}}{D}\right) - \sin\left(\frac{2k\pi x_i}{D}\right)\right)\right] T_k(t) \\ + \sum_{\substack{n=0 \\ n \neq k}}^{\infty} \frac{\alpha_i}{2} \left[\begin{array}{l} \frac{D}{(n+k)\pi}\left(\sin\left(\frac{(n+k)\pi x_{i+1}}{D}\right) - \sin\left(\frac{(n+k)\pi x_i}{D}\right)\right) \\ + \frac{D}{(n-k)\pi}\left(\sin\left(\frac{(n-k)\pi x_{i+1}}{D}\right) - \sin\left(\frac{(n-k)\pi x_i}{D}\right)\right) \end{array} \right] T_n(t) \end{array} \right\}$$

for $k = 0$ $\hspace{4cm}$ (2.33)

$$D\frac{\partial T_k(t)}{\partial t} = \alpha_{m-1} \left. \frac{\partial T(x,t)}{\partial x} \right|_{x=x_m^-} - \alpha_1 \left. \frac{\partial T(x,t)}{\partial x} \right|_{x=x_1}$$

$$-\frac{n\pi}{D} \sum_{i=1}^{m-2} \left\{ \sum_{n=1}^{\infty} (\alpha_i - C_{i,i+1}\alpha_{i+1}) \sin\left(\frac{n\pi x_{i+1}}{D}\right) T_n(t) \right\}$$

where k^{th} and n^{th} are subscripts that refer to the index of the specific harmonic and i^{th} subscript refers to the material layer.

The temperature gradient at the boundaries of the die and package can be determined by:

$$\left. \frac{\partial T(x,t)}{\partial x} \right|_{x_1} = -\frac{q(t)}{k_1}$$

$$\left. \frac{\partial T(x,t)}{\partial x} \right|_{x_m^-} = -\frac{h(T(x_m,t) - T_0)}{k_{m-1}}$$

$\hspace{10cm}$ (2.34)

where $q(t)$ is the heat dissipation function produced by internal energy losses in the semiconductor, k_1 and k_{m-1} are the thermal conductivity of the first and the last layer, and h is the convective heat transfer coefficient of the heatsink. Then the 1-D heat conduction equation, reduced to the form of Eq. (2.33), is comprised of a set of first-order differential equations. The unknown time-dependent harmonic amplitudes can be found using the numerical simulation described in the following chapter. The temperature at the two ends of the DBC can be calculated by the following equations:

$$T(x_1, t) = \sum_{k=0}^{\infty} T_k(t)$$

$$T(x_m, t) = \sum_{k=0}^{\infty} (-1)^k T_k(t)$$

(2.35)

where k is a subscript that refers to the index of the specific harmonic.

CHAPTER 3

Realization of Power IGBT and Diode Thermal Model

3.1 INTRODUCTION

Level-3 models for a power diode and IGBT, under inductive load switching, using a Fourier-based solution of the ADE describing the lightly doped N^- drift regions were developed and are presented in *Modeling Bipolar Power Semiconductor Devices*. The electrical circuit schematic of an IGBT using an inductive load and free-wheeling diode is shown in Fig. 3.1. The circuit was simulated in Matlab and Simulink using diode and IGBT electro-thermal models.

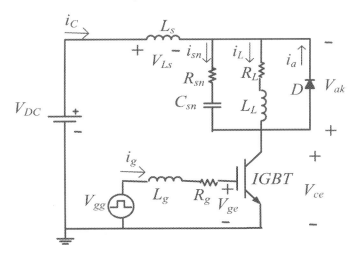

Figure 3.1: IGBT circuit with an inductive load and free-wheeling diode, D.

For a complete simulation model, the thermal issue cannot be neglected since most of the primary parameters of a semiconductor material are strongly temperature dependent. The modification of semiconductor material parameters should be included based on thermal feedback. A schematic description of the complete electro-thermal simulation model is illustrated in Fig. 3.2.

The electrical and thermal models are connected through an estimation of self-heat due to dissipation. Figure 3.3 shows the interconnection between the electrical and thermal models for the diode and IGBT in Simulink.

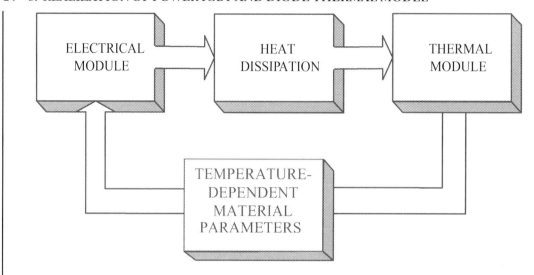

Figure 3.2: Basic diagram of complete electro-thermal simulation model.

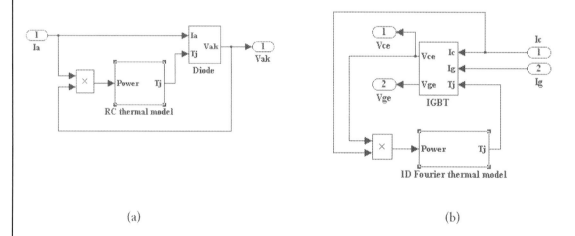

(a)   (b)

Figure 3.3: (a) Diode model and (b) IGBT model.

Two different methods are used to calculate the junction temperatures of the diode and IGBT. An equivalent *RC* network is used to calculate the diode junction temperature, and a one-dimensional Fourier series is used to calculate the IGBT junction temperature. The use of an *RC* thermal model simplifies the simulation and is accurate enough for the diode (device not under test). The Fourier description for the thermal model is more accurate than the *RC* model and is employed with the IGBT since it is the device of interest in the simulation. Once the junction temperatures are calculated, they are used to determine the temperature-dependent parameter values discussed in Chapter 1, intrinsic carrier concentration, ionized impurity concentration, carrier mobility, electron and hole diffusivities, ambipolar diffusivity, carrier lifetime, emitter recombination parameter, threshold voltage, and transconductance.

The parameters are recalculated, based on the updated device temperature, and used in the next time-step of the electrical simulation. Since the thermal response is slower compared to electrical behavior, it is reasonable to employ the thermal feedback in discrete time steps larger than the time-steps used in the electrical simulation without any impact on the results. It has been found that only one out of ten samples of the calculated temperature needs to be fed to the electrical simulation model, as diagrammed in Fig. 3.4. This will speed up the overall simulation process without sacrificing accuracy.

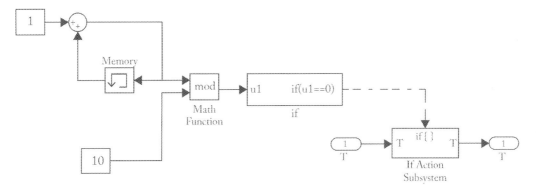

Figure 3.4: Simulink diagram of discrete modification of material parameters of the semiconductors.

3.2   REALIZATION OF EQUIVALENT *RC* NETWORK

The junction temperature of the diode is calculated by an equivalent *RC* network. The number of *RC* cells is equal to the number of layers used for the device package. The thermal resistance, *R*, and thermal capacitance, *C*, depend on the geometries and the properties of each corresponding layer. These values are calculated using Eqs. (2.10)–(2.12). The heat transfer, *q*, and the temperature, *T*, from the thermal equivalent circuit can be calculated in a way similar to the current, *i*, and potential, *v*, in an electrical circuit. A part of a *RC* thermal circuit is presented in Fig. 3.5(a).

Kirchoff's current law is used to describe the circuit. For the node *i*, the equation is:

$$q_{C_{i-1}} = q_{i-1} - q_i .$$ (3.1)

The heat flow, q_i, through the thermal resistor, R_i, is calculated using Ohm's law:

$$q_i = \frac{T_i - T_{i+1}}{R_i} .$$ (3.2)

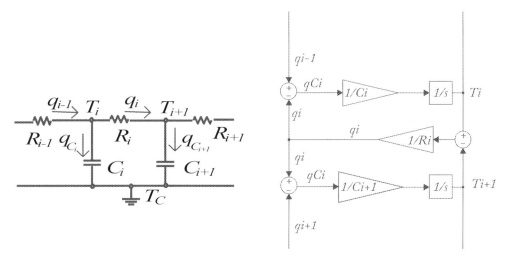

Figure 3.5: A part of a RC thermal circuit with constant thermal conductivity, k_i, and constant heat capacity, c_i: (a) circuit diagram and (b) realization of the diagram in Simulink.

The temperature difference across the two nodes of the capacitor C_i is calculated by:

$$T_i - T_C = \frac{1}{C_i} \int q_{C_i} \, dt \, . \tag{3.3}$$

The implementation of the RC thermal network with constant thermal conductivity, k_i, and constant specific heat capacity, c_i, in Simulink is shown in Fig. 3.5(b). For implementation of Eq. (3.3), an integrator block is used with an initial condition T_C.

The diagram of an equivalent RC thermal network, with constant thermal conductivity, k_i, and specific heat capacity, c_i, of a packaged diode realized in Simulink, is presented in Fig. 3.6. The thermal conductivity, k_i, and the specific heat capacity, c_i, of the different layers of the package are assumed to be constant with respect to temperature. Table 3.1 lists the seven material layers modeled in Fig. 3.6.

As it was mentioned in Section 2.3, including the temperature dependency of the thermal conductivity and heat capacity in the model will decrease the error during high dissipation conditions. Figure 3.7 shows a Simulink implementation of the RC thermal network with non-constant thermal conductivity, k_i, and non-constant specific heat capacity, c_i. The diagram consists of seven subsystems, equal to the number of package layer materials. Each subsystem calculates the temperature at the top of the corresponding layer and heat flow passing through the layer. A Simulink diagram of the silicon layer subsystem is presented in Fig. 3.8. At each simulation step, the subsystem calculates the thermal conductivity, k_i, and specific heat capacity, c_i, of the layer using the temperature of the top of the layer, by utilizing the least squares regression fit coefficients in Table 2.1. Then, Eqs. (2.10)–(2.12) are used for calculating the updated values of conduction thermal

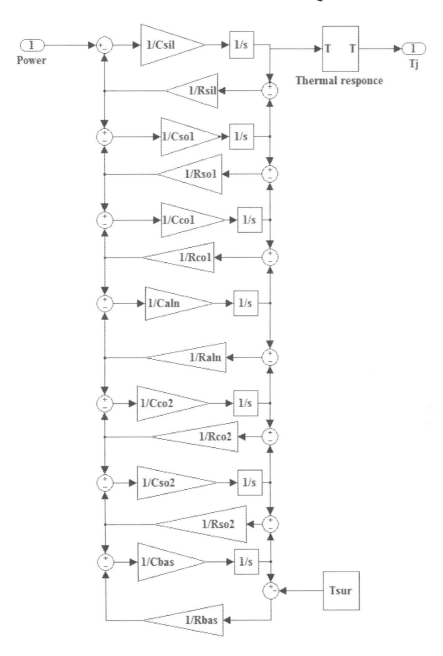

Figure 3.6: Implementation of an *RC* thermal circuit with constant thermal conductivity, k_i, and specific heat capacity, c_i, in a Simulink diagram.

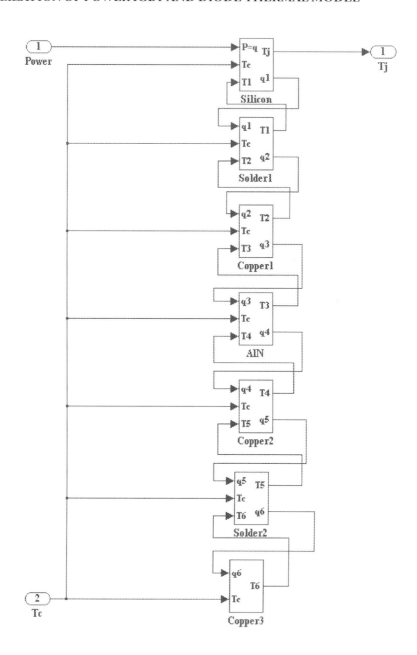

Figure 3.7: Implementation of an RC thermal circuit with non-constant thermal conductivity, k_i, and non-constant specific heat capacity, c_i, in a Simulink diagram.

Table 3.1: List of material layers

Name	Description
sil	Silicon junction
sol	Solder 1
col	Copper 1
aln	Aluminum Nitride
co2	Copper 2
so2	Solder 2
bas	Baseplate

resistance,R_i, and of thermal capacitance, C_i. Equations (3.2) and (3.3) are used for calculating the heat flow passing through the layer and the temperature at the top of the layer.

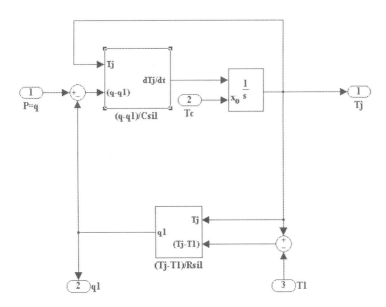

Figure 3.8: Implementation of Silicon subsystem from Fig. 3.7.

3.3   REALIZATION OF ONE-DIMENSIONAL FOURIER-SERIES THERMAL MODEL

To calculate the junction temperature of a device, the dissipated heat is used as the input parameter. The solution of the heat diffusion equation, Eq. (2.15), is derived in Chapter 2. It can be rewritten as:

for $k \neq 0$:

$$
\frac{\partial T_k(t)}{\partial t} + \frac{2\pi}{D^2} \sum_{i=1}^{m-2} \left\{ \sum_{n=1}^{\infty} \left[\begin{array}{l} n\,(\alpha_i - C_{i,i+1}\alpha_{i+1}) \sin\left(\frac{n\pi x_{i+1}}{D}\right) \cos\left(\frac{k\pi x_{i+1}}{D}\right) \\ -k\,(\alpha_i - \alpha_{i+1}) \sin\left(\frac{k\pi x_{i+1}}{D}\right) \cos\left(\frac{n\pi x_{i+1}}{D}\right) \end{array} \right] T_n(t) \right\}
$$

$$
+ \frac{2}{D}\left(\frac{k\pi}{D}\right)^2 \sum_{i=1}^{m-1} \left\{ \begin{array}{l} \frac{\alpha_i}{2}\left[x_{i+1} - x_i + \frac{D}{2k\pi}\left(\sin\left(\frac{2k\pi x_{i+1}}{D}\right) - \sin\left(\frac{2k\pi x_i}{D}\right)\right)\right] T_k(t) \\[2mm] + \sum_{\substack{n=0 \\ n\neq k}}^{\infty} \frac{\alpha_i}{2}\left[\begin{array}{l} \frac{D}{(n+k)\pi}\left(\sin\left(\frac{(n+k)\pi x_{i+1}}{D}\right) - \sin\left(\frac{(n+k)\pi x_i}{D}\right)\right) \\ + \frac{D}{(n-k)\pi}\left(\sin\left(\frac{(n-k)\pi x_{i+1}}{D}\right) - \sin\left(\frac{(n-k)\pi x_i}{D}\right)\right) \end{array} \right] T_n(t) \end{array} \right\}
$$

$$
- \frac{2\alpha_{m-1}}{D}(-1)^k \left.\frac{\partial T(x,t)}{\partial x}\right|_{x=x_m^-}
$$

$$
= -\frac{2\alpha_1}{D} \left.\frac{\partial T(x,t)}{\partial x}\right|_{x=x_1}
$$

for $k = 0$

$$
\frac{\partial T_k}{\partial t} + \sum_{i=1}^{m-2} \left\{ \sum_{n=1}^{\infty} \frac{n\pi}{D^2}(\alpha_i - C_{i,i+1}\alpha_{i+1}) \sin\left(\frac{n\pi x_{i+1}}{D}\right) T_j(t) \right\} - \frac{\alpha_{m-1}}{D} \left.\frac{\partial T(x,t)}{\partial x}\right|_{x=x_m^-}
$$

$$
= -\frac{\alpha_1}{D} \left.\frac{\partial T(x,t)}{\partial x}\right|_{x=x_1} \tag{3.4}
$$

where k^{th} and n^{th} subscripts refer to the index of the specific harmonic and the ith subscript refers to the material layer.

The junction temperature is:

$$
T(x_1,t) = \sum_{k=0}^{\infty} T_k(t) . \tag{3.5}
$$

From the boundary conditions, Eq. (2.34), it can be seen that the third term on the left side of Eq. (3.4) is a constant. If the solution of the harmonic, T_k, is substituted in Eq. (3.5), this second term will cancel out. Using the boundary conditions, the solution Eq. (3.4) can be rewritten as:

$$
\frac{d\bar{T}}{dt} + A \times \bar{T} = B \tag{3.6}
$$

where $\bar{T}$ is a $(k+1)$ dimensional vector of the amplitudes of the harmonics $[T_0, T_1, \ldots T_k]$, each term of which is a function of time. A is a transfer matrix. The matrix has $(k + 1)$ rows, because that is the number of harmonics and $(k + 1)$ columns, because that is the number of equations that will be used. B is an input matrix with one column and $(k + 1)$ rows and it represents the heat generation term. After some manipulation of the matrices, the elements of A and B are determined by the following equations.

The matrix A is:

$$A = \begin{pmatrix} a_{00} & a_{01} & \ldots & a_{0q} & \ldots & a_{0k} \\ a_{10} & a_{11} & \ldots & a_{1q} & \ldots & a_{1k} \\ \vdots & \vdots & & \vdots & & \vdots \\ a_{p0} & a_{p1} & \ldots & a_{pq} & \ldots & a_{pk} \\ \vdots & \vdots & & \vdots & & \vdots \\ a_{k0} & a_{k1} & \ldots & a_{kp} & \ldots & a_{kk} \end{pmatrix}. \tag{3.7}$$

Using the temperature gradient at the boundaries of the die and package, Eq. (2.34), and the temperature of the end of the DBC with coordinate x_m, Eq. (2.35), the elements of the matrix A can be determined using the following equations.

For $p = 0$ and $q = 0 \ldots k$

$$a_{0q} = \frac{\alpha_{m-1} h (-1)^q}{D k_{m-1}} + \frac{q\pi}{D^2} \sum_{i=1}^{m-2} (\alpha_i - C_{i,i+1}\alpha_{i+1}) \sin\left[\frac{q\pi x_{i+1}}{D}\right]. \tag{3.8}$$

For $p = 1 \ldots k$ and $q = 0$

$$a_{p0} = \frac{-2\alpha_{m-1} h (-1)^p}{D k_{m-1}}. \tag{3.9}$$

For $p = q = 1 \ldots k$

$$a_{pp} = -\frac{2\alpha_{m-1}h}{D k_{m-1}} + \frac{p\pi}{D^2} \left(\sum_{i=1}^{m-2} (\alpha_i - C_{i,i+1}\alpha_{i+1}) \sin\left[\frac{2p\pi x_{i+1}}{D}\right] - (\alpha_i - \alpha_{i+1}) \sin\left[\frac{2p\pi x_{i+1}}{D}\right] \right)$$

$$- \frac{(p\pi)^2}{D^3} \sum_{i=1}^{m-1} \alpha_1 (x_{i+1} - x_i) + \frac{p\pi}{2D^2} \sum_{i=1}^{m-1} (\alpha_i - \alpha_{i+1}) \left[\sin\left[\frac{2p\pi x_{i+1}}{D}\right] - \sin\left[\frac{2p\pi x_i}{D}\right] \right]$$

$$= -\frac{2\alpha_{m-1}h}{D k_{m-1}} + \frac{p\pi}{D^2} \sum_{i=1}^{m-2} (\alpha_{i+1} - C_{i,i+1}\alpha_{i+1}) \sin\left[\frac{2p\pi x_{i+1}}{D}\right] - \frac{(p\pi)^2}{D^3} \sum_{i=1}^{m-1} \alpha_1 (x_{i+1} - x_i)$$

$$+ \frac{p\pi}{2D^2} \sum_{i=1}^{m-1} (\alpha_i - \alpha_{i+1}) \left[\sin\left[\frac{2p\pi x_{i+1}}{D}\right] - \sin\left[\frac{2p\pi x_i}{D}\right] \right]. \tag{3.10}$$

For $p = 1 \ldots k, q = 1 \ldots q, p \neq q$

$$a_{pq} = \frac{2\alpha_{m-1}h(-1)^{p+q}}{Dk_{m-1}} - \frac{2\pi}{D^2} \sum_{i=1}^{m-1} \sum_{q=1}^{\infty} \left[q \left(\alpha_i - C_{i,i+1}\alpha_{i+1}\right) \sin\left(\frac{q\pi x_{i+1}}{D}\right) \cos\left(\frac{p\pi x_{i+1}}{D}\right) \right.$$

$$\left. - p \left(\alpha_i - \alpha_{i+1}\right) \sin\left(\frac{p\pi x_{i+1}}{D}\right) \cos\left(\frac{q\pi x_{i+1}}{D}\right) \right]$$

$$+ \frac{\pi p^2}{D^2} \sum_{i=1}^{m-1} \sum_{\substack{q=0 \\ p \neq q}}^{\infty} \alpha_i \left[\frac{1}{(q+p)} \left(\sin\left(\frac{(q+p)\pi x_{i+1}}{D}\right) - \sin\left(\frac{(q+p)\pi x_i}{D}\right) \right) \right.$$

$$\left. + \frac{1}{(q-p)} \left(\sin\left(\frac{(q-p)\pi x_{i+1}}{D}\right) - \sin\left(\frac{(q-p)\pi x_i}{D}\right) \right) \right]. \qquad (3.11)$$

The matrix B is:

$$B = \begin{pmatrix} b_0 \\ b_1 \\ \vdots \\ b_k \end{pmatrix}. \qquad (3.12)$$

The terms $b_1, b_2, \ldots, b_k$ are the collection of derivative terms evaluated at $x = x_1$ on the right side of Eq. (3.4) after substituting for the relations in Eq. (2.34).

$$b_0 = \frac{\alpha_1}{Dk_1} q(t)$$

$$b_1 = b_2 = \cdots = b_k = \frac{2\alpha_1}{Dk_1} q(t) \qquad (3.13)$$

where $q(t)$ is the self-heating generation function in the silicon chip, which can be obtained from the electrical circuit simulation. Using this formulation, the partial differential equation, Eq. (2.13), has been transformed into a set of first-order ordinary differential equations that can easily be solved, Eq. (3.4), to give the time-evolution of the temperature at each position in the packaged device or module.

The Fourier series thermal model is based on the basic physics equations of heat conduction. The computational speed of the model depends on the number of harmonics used in the description. Compared with an electrical model based on the Fourier series expansion, a stationary boundary location will simplify the solution process. However, because of the existence of multiple layers, the total number of harmonics is higher than that required in the electrical model.

The implementation of a Fourier-based thermal model in Simulink is shown in Fig. 3.9. It calculates the vector of harmonic amplitudes, $\bar{T}$, using Eq. (3.6). The model further calculates the semiconductor junction temperature using Eq. (3.5) in the T_j subsystem block. Due to the relatively large value for the thermal time constants as compared to the electrical time constants, the electrical model updates the semiconductor material parameters once every ten time steps.

The subsystems given in Figs. 3.6, 3.7, and 3.9 can be used to calculate the junction temperature of a power semiconductor device during operation. By comparing the changes in the junction temperature during the operation of a new device and the working device, the reliability of the device can be predicted.

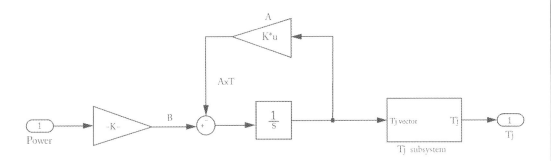

Figure 3.9: Implementation of Fourier-based thermo model in Simulink.

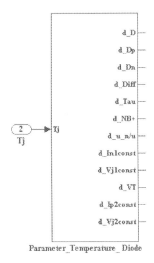

Figure 3.10: A Simulink subsystem for calculating temperature-dependent parameters of the diode.

3.4   TEMPERATURE DEPENDENT PARAMETERS OF DIODES AND THEIR CONNECTION TO AN ELECTRICAL MODEL

The diode temperature-dependent parameters are calculated by the equations given in Chapter 1, Section 1.2. A Parameter Temperature Diode (PTD) subsystem is shown in Fig. 3.10. It has an input, the diode junction temperature T_j, and 12 outputs, the calculated temperature-dependent material parameters. The updated parameters are then used by the diode electrical model. The realization of PTD subsystem in Simulink is presented in Fig. 3.11.

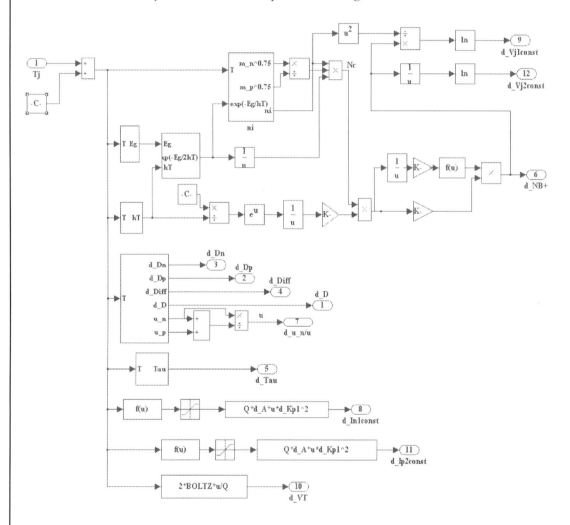

Figure 3.11: Realization of PTD subsystem in Simulink.

A model of a power diode using a Fourier-based solution of the ambipolar diffusion equation (ADE) in the lightly doped N^- drift region, including the temperature-dependent parameters calculated at every iteration, is presented below. The P^+NN^+ diode has three distinct regions. The basic equations governing the behavior of semiconductor devices are used to analyze the operation of the diode. The diode current, I_D, the electron, hole, and displacement currents at junctions J_1 and J_2 (I_{n1}, I_{p1}, I_{disp1} and I_{n2}, I_{p2}, I_{disp2}), and their direction are shown in Fig. 3.12.

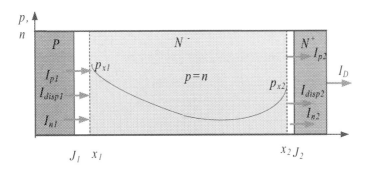

Figure 3.12: Schematic structure of a $P^+N^-N^+$ power diode during the on state.

Figure 3.13 presents the connections of the 12 outputs of the PTD subsystem with the three electrical model subsystem implemented in Simulink.

Referring to the PTD subsystem shown in Fig. 3.11, from the calculated junction temperature T_j, the electron and hole mobilities μ_n and μ_p are calculated in the PTD subsystem by Eq. (1.8) and used to calculate the electron and hole diffusivities D_n and D_p using Eq. (1.9). Then the ambipolar diffusivity D and diffusivity D_{diff} are calculated using Eqs. (1.10) and (1.11), respectively. The carrier lifetime τ is calculated using Eq. (1.12). The electron and hole density of states effective masses, m_n^* and m_p^*, are calculated using Eqs. (1.2) and (1.3). The energy band gap, E_G, is calculated by Eq. (1.4) and the effective density of states, N_C, by Eq. (1.6). The next thermally dependent parameter calculated by the PTD subsystem is the ionized impurity concentration in the lightly doped drift region, $N_{N^-}^+$ using Eq. (1.5). The ratio given by Eq. (3.14) is also calculated in the PTD subsystem.

$$\mu = \frac{\mu_n}{\mu_n + \mu_p} . \tag{3.14}$$

The temperature-dependent parameters calculated, D, D_n, D_p, D_{diff}, τ, Q_N, and μ, are denoted as d_D, d_Dn, d_Dp, d_Diff, d_Tau, d_NB+, and d_u_n/u, as shown in Figure 3.13. These parameters are inputs to the N^- drift subsystem of the power diode. Figure 3.14 shows the N^- drift region subsystem, including the temperature-dependent parameters, implemented in Simulink. It can be seen that the temperature-dependent parameters are input only to Carrier

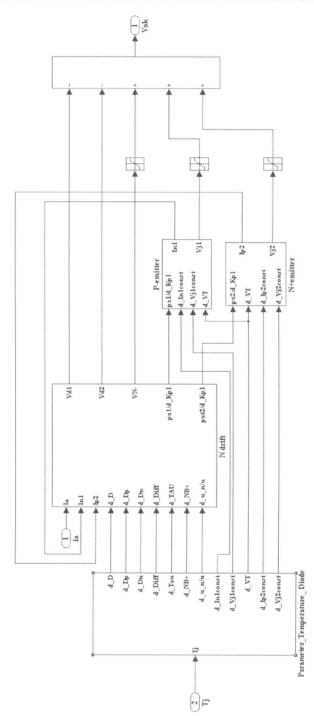

Figure 3.13: PTD subsystem and the three diode electrical model subsystems in Simulink.

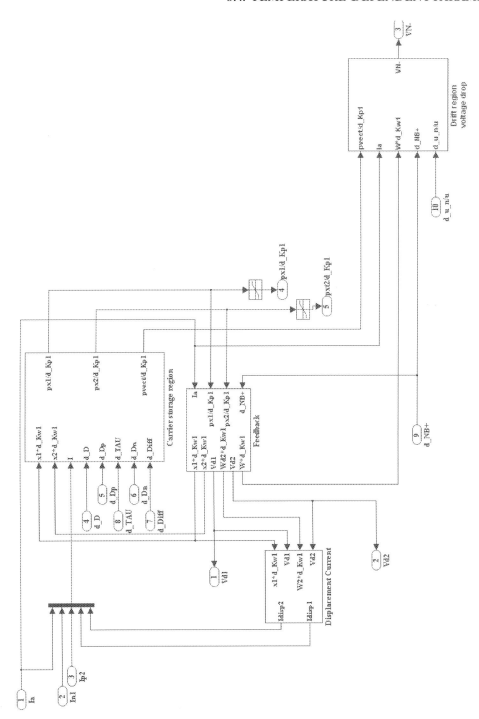

Figure 3.14: The N^- drift region subsystem with temperature-dependent parameters implemented in Simulink.

storage, Feedback, and Drift region voltage drop sub-subsystems. The displacement current sub-subsystem is realized by equations that do not include temperature-dependent parameters.

As shown in Fig. 3.14, the Carrier storage region sub-subsystem has the following temperature-dependent parameters as inputs: d_D, d_Dn, d_Dp, d_Diff, and d_Tau. The subsystem provides the solution to the ADE, Eq. (3.15), by using the Fourier solution Eq. (3.16) and the boundary conditions of Eq. (3.17).

$$D_{N-}\frac{\partial^2 p(x,t)}{\partial x^2} = \frac{p(x,t)}{\tau_{N-}} + \frac{\partial p(x,t)}{\partial t} \tag{3.15}$$

for $k > 0$:

$$D_{N-}\left[\frac{\partial p(x,t)}{\partial x}\bigg|_{x_2}(-1)^k - \frac{\partial p(x,t)}{\partial x}\bigg|_{x_1}\right] = \frac{x_2 - x_1}{2}\left(\frac{d\,p_k(t)}{dt} + \left[\frac{1}{\tau} + \frac{D_{N-}k^2\pi^2}{(x_2 - x_1)^2}\right]p_k(t)\right)$$

$$+ \left(\sum_{\substack{n=1 \\ n \neq k}}^{\infty}\frac{n^2}{n^2 - k^2}\left[\frac{d\,x_1}{dt} - (-1)^{n+k}\frac{d\,x_2}{dt}\right]p_n(t) + \frac{p_k(t)}{4}\left(\frac{d\,x_1}{dt} - \frac{d\,x_2}{dt}\right)\right)$$

for $k = 0$ \hfill (3.16)

$$D_{N-}\left[\frac{\partial p(x,t)}{\partial x}\bigg|_{x_2} - \frac{\partial p(x,t)}{\partial x}\bigg|_{x_1}\right] = (x_2 - x_1)\left(\frac{d\,p_0(t)}{dt} + \frac{p_0(t)}{\tau}\right)$$

$$+ \sum_{n=1}^{\infty}\left[\frac{d\,x_1}{dt} - (-1)^n\frac{d\,x_2}{dt}\right]p_n(t).$$

The required boundary conditions to solve for the carrier densities p_{x1} and p_{x2} are: the boundaries of the region (x_1 and x_2), their time differentials (dx_1/dt and dx_2/dt), and the carrier density gradients at x_1 and x_2 as given by the Eqs. (3.17):

$$\frac{\partial p}{\partial x}\bigg|_{x_1} = \frac{1}{2q}\left(\frac{J_n}{D_n} - \frac{J_p}{D_p}\right)\bigg|_{x_1} \quad \text{and} \quad \frac{\partial p}{\partial x}\bigg|_{x_2} = \frac{1}{2q}\left(\frac{J_n}{D_n} - \frac{J_p}{D_p}\right)\bigg|_{x_2}. \tag{3.17}$$

The current continuity equation at junction J_2 is:

$$I_{n2} = I_D - I_{p2} - I_{disp2}. \tag{3.18}$$

Assuming that:

$$\frac{1}{D_n} + \frac{1}{D_p} = \frac{D_n + D_p}{D_n D_p} = \frac{2}{D_{N-}} = \frac{2}{D} \quad \text{and}$$

$$\frac{D}{2}\left(\frac{1}{D_p} - \frac{1}{D_n}\right) = \frac{D_n - D_p}{D_n + D_p} = D_{diff}, \tag{3.19}$$

and using the current continuity equations at junctions J_1, Eq. (3.20):

$$I_D = I_{n1} + I_{p1} + I_{disp1} \tag{3.20}$$

and J_2, Eq. (3.21):

$$I_{n2} = I_D - I_{p2} - I_{disp2} \tag{3.21}$$

then, the left side of Eqs. (3.16) for even and odd numbers of k can be calculated by the following equations:

$$I_{even} = D \left(\left.\frac{\partial p}{\partial x}\right|_{x_2} - \left.\frac{\partial p}{\partial x}\right|_{x_1} \right) = \begin{vmatrix} I_D \\ I_{p2} \\ I_{n1} \\ I_{disp1} \\ I_{disp2} \end{vmatrix} \frac{1}{2qA} \begin{vmatrix} 2 & -2 & -2 & -\dfrac{D}{D_p} & -\dfrac{D}{D_n} \end{vmatrix} \tag{3.22}$$

$$I_{odd} = -D \left(\left.\frac{\partial p}{\partial x}\right|_{x_2} + \left.\frac{\partial p}{\partial x}\right|_{x_1} \right) = \begin{vmatrix} I_D \\ I_{p2} \\ I_{n1} \\ I_{disp1} \\ I_{disp2} \end{vmatrix} \frac{1}{2qA} \begin{vmatrix} -2D_{diff} & 2 & 2 & -\dfrac{D}{D_p} & \dfrac{D}{D_n} \end{vmatrix} .$$

When no thermal model is implemented, a Matlab program precalculates a constant matrix; (d_boudmax2) and the Simulink program uses it during the simulation. A more in-depth discussion of the electrical model is provided in [7]. For a thermal model, since the parameters d_D, d_Dn, d_Dp, and d_Diff are temperature dependent; the matrix should be recalculated at every iteration. For example, the matrix d_boudmat2 with six harmonics is:

$$\text{d\_boudmat2} = \frac{1}{2qA} \begin{vmatrix} 2 & -2 & -2 & -\dfrac{D}{D_p} & -\dfrac{D}{D_n} \\ -2D_{diff} & -2 & 2 & -\dfrac{D}{D_p} & \dfrac{D}{D_n} \\ 2 & -2 & -2 & -\dfrac{D}{D_p} & -\dfrac{D}{D_n} \\ -2D_{diff} & -2 & 2 & -\dfrac{D}{D_p} & \dfrac{D}{D_n} \\ 2 & -2 & -2 & -\dfrac{D}{D_p} & -\dfrac{D}{D_n} \\ -2D_{diff} & -2 & 2 & -\dfrac{D}{D_p} & \dfrac{D}{D_n} \\ 2 & -2 & -2 & -\dfrac{D}{D_p} & -\dfrac{D}{D_n} \end{vmatrix} . \tag{3.23}$$

Since the parameters d_D, d_Dn, d_Dp, and d_Diff are temperature dependent, the matrix d_boudmat2 in Simulink needs to be recalculated at every time step during the simulation. In this case, the matrix can be presented and implemented in Simulink as a sum of four matrices multiplied by different constants:

$$d\_boudmat2 = \frac{1}{2qA} d\_Kp1 \left[d\_b1 - D_{diff} * d\_b2 + \frac{D}{D_p} * d\_b3 + \frac{D}{D_n} * d\_b4 \right] \quad (3.24)$$

where d_Kp1=10^{15} is a scaling factor used to help the simulation converge, and the four matrices d_b1, d_b2, d_b3, and d_b4 are:

$$d\_b1 = \begin{vmatrix} 2 & -2 & -2 & 0 & 0 \\ 0 & -2 & 2 & 0 & 0 \\ 2 & -2 & -2 & 0 & 0 \\ 2 & -2 & -2 & 0 & 0 \\ 0 & -2 & 2 & 0 & 0 \\ 2 & -2 & -2 & 0 & 0 \\ 2 & -2 & -2 & 0 & 0 \end{vmatrix}, \quad d\_b2 = \begin{vmatrix} 0 & 0 & 0 & 0 & 0 \\ 2 & 0 & 0 & 0 & 0 \\ 0 & 0 & 0 & 0 & 0 \\ 2 & 0 & 0 & 0 & 0 \\ 0 & 0 & 0 & 0 & 0 \\ 2 & 0 & 0 & 0 & 0 \\ 0 & 0 & 0 & 0 & 0 \end{vmatrix},$$

$$d\_b3 = \begin{vmatrix} 0 & 0 & 0 & -1 & 0 \\ 0 & 0 & 0 & -1 & 0 \\ 0 & 0 & 0 & -1 & 0 \\ 0 & 0 & 0 & -1 & 0 \\ 0 & 0 & 0 & -1 & 0 \\ 0 & 0 & 0 & -1 & 0 \\ 0 & 0 & 0 & -1 & 0 \end{vmatrix}, \quad d\_b4 = \begin{vmatrix} 0 & 0 & 0 & 0 & -1 \\ 0 & 0 & 0 & 0 & 1 \\ 0 & 0 & 0 & 0 & -1 \\ 0 & 0 & 0 & 0 & 1 \\ 0 & 0 & 0 & 0 & -1 \\ 0 & 0 & 0 & 0 & 1 \\ 0 & 0 & 0 & 0 & -1 \end{vmatrix}.$$

In Fig. 3.15, the Carrier storage subsystem, including the thermal-dependent parameters, is presented. The main difference between the two carrier storage subsystems, one including and the other not including the temperature-dependent parameters, is that the carrier storage subsystem for the thermal model has an additional subsystem d_boudmat2 that is used to calculate the matrix given by Eq. (3.24). A sub-subsystem calculating the d_boudmat2 matrix in Simulink is presented in Fig. 3.16.

The N^- drift region subsystem, shown in Fig. 3.14, has another sub-subsystem, Feedback, which has a temperature-dependent parameter as an input. This parameter is the ionized impurity concentration of the lightly doped drift region N_{N-}^+. The implementation of the Feedback subsystem is shown in Fig. 3.17. The Feedback sub-subsystem includes a Depletion Layer Widths sub-sub-subsystem that uses the ionized impurity concentration of the lightly doped drift region N_{N-}^+ to calculate the depletion widths W_{d1} and W_{d2}:

$$W_{d1} = \sqrt{\frac{2\varepsilon V_{d1}}{qN_{N-}^+ + \frac{|I_{p1}|}{Av_{sat}}}}$$

$$(3.25)$$

$$W_{d2} = \sqrt{\frac{2\varepsilon V_{d2}}{qN_{N-}^+ + \frac{|I_{n2}|}{Av_{sat}}}}.$$

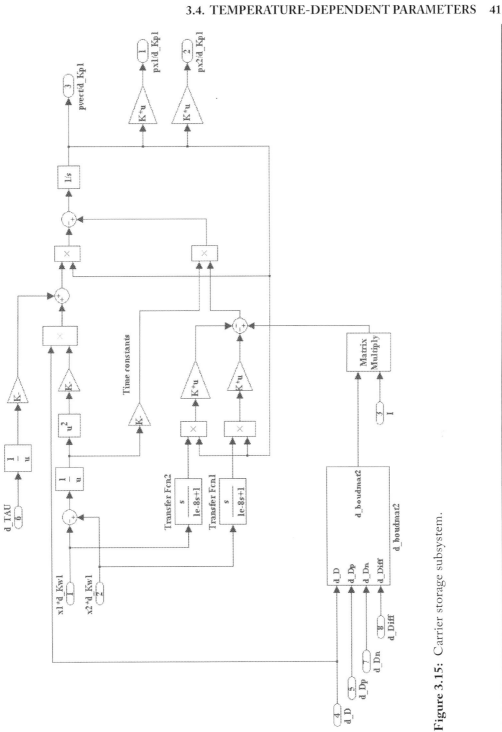

Figure 3.15: Carrier storage subsystem.

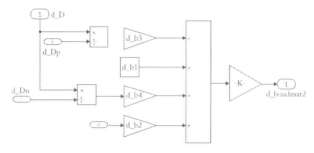

Figure 3.16: Matrix d_boudmat2 realized in Simulink.

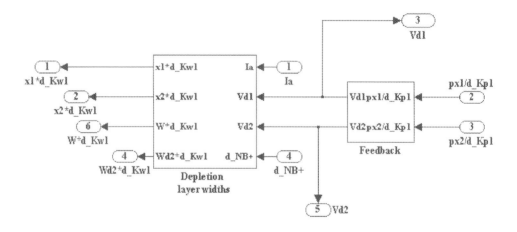

Figure 3.17: Feedback sub-subsystem.

The implementation of the Depletion Layer Widths sub-sub-subsystem is shown in Fig. 3.18.

The Drift region voltage drop sub-subsystem, shown in Fig. 3.14, has two temperature-dependent parameters, the ionized impurity concentration in the lightly doped drift region N_{N-}^{+} (d_Nb) and the mobility ratio calculated by Eq. (3.14) (d_u_n/u). The voltage drop in the lightly doped drift region, V_{N-}, is calculated based on the injected carrier concentration.

$$V_{N-} \approx \frac{I_C}{qA(\mu_n + \mu_p)} \frac{x_2 - x_1}{-1}$$

$$\sum_{k=0}^{M-1}\left[\frac{1}{p_T(k) - p_T(k-1)} \ln\left(\frac{p_T(k)}{p_T(k-1)}\right)\right] + V_T\left(\frac{\mu_n - \mu_p}{\mu_n + \mu_p}\right)\ln\left(\frac{p_{x2}}{p_{x1}}\right) \quad (3.26)$$

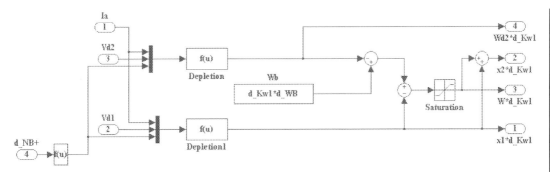

Figure 3.18: Depletion layer widths sub-sub-subsystem.

where the carrier distribution $p_T(k)$ is calculated as:

$$p_T(k) = p\left(x_1 + \frac{k\,(x_2 - x_1)}{M - 1}\right) + \frac{\mu_n N_{N^-}}{\mu_n + \mu_p}. \tag{3.27}$$

The Drift region voltage drop sub-subsystem realized in Simulink is shown in Fig. 3.19.

The other two regions of the power diode, P and N^+ emitters, are modeled by the two subsystems, P-emitter and N^+-emitter. These subsystems are used to calculate the minority carrier currents and junction voltages.

Emitter recombination parameters, h_n and h_p, are used in the relations for calculation of the injected currents into the drift region (N^- base):

$$I_{n1} = qAh_p\,p_{x_1}^2 \tag{3.28}$$

$$I_{p2} = qAh_n\,p_{x_2}^2 \tag{3.29}$$

where p_{x_1} and p_{x_2} are the excess carrier concentrations at junction J_1 and J_2, respectively (see Fig. 3.12).

Since the emitter recombination parameters h_n and h_p depend on temperature, they are calculated at every iteration in the PTD subsystem; the two expressions, d_In1const and d_Ip2const, are then used for calculation of the minority currents in the two regions based on Eqs. (3.28) and (3.29). The two currents are calculated by the following equations:

$$d\_In1const = qAh_p\,d\_Kp1^2 \tag{3.30}$$

$$d\_Ip2const = qAh_n\,d\_Kp1^2 . \tag{3.31}$$

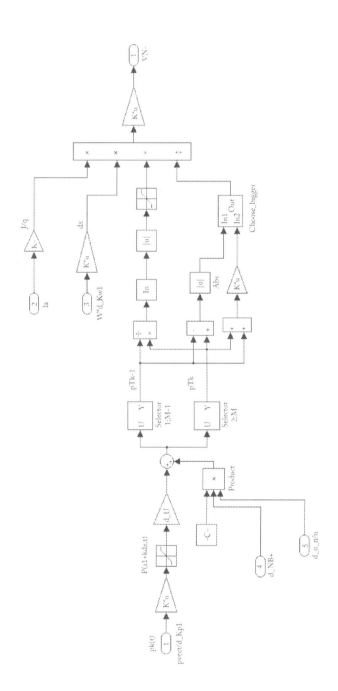

Figure 3.19: The Drift region voltage drop sub-subsystem.

The junction voltages across J_1 and J_2 are calculated as:

$$V_{j1} = V_T \ln \left(\frac{p_{x1} N_{N-}^+}{n_i^2} \right) , \qquad (3.32)$$

$$V_{j2} = V_T \ln \left(\frac{p_{x2}}{N_{N-}} \right) . \qquad (3.33)$$

Three constants, d_VT, d_Vj1const, and d_Vj2const, are calculated in the PTD subsystem and used for calculation of the two junction voltages. The constants are calculated by the following equations:

$$d\_VT = \frac{2kT}{q} , \qquad (3.34)$$

$$d\_Vj1const = \ln \left(\frac{N_{N-}^+}{n_i^2} \right) , \qquad (3.35)$$

$$V_{j2} = \ln \left(\frac{1}{N_{N-}^+} \right) . \qquad (3.36)$$

The realization of the P-emitter subsystem is shown in Figure 3.20.

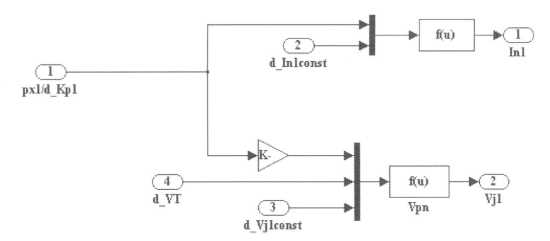

Figure 3.20: P-emitter subsystem.

Similarly, the realization of the N^+ emitter subsystem is shown in Figure 3.21.

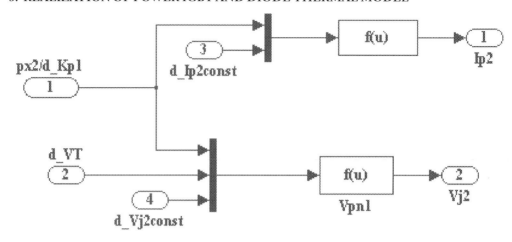

Figure 3.21: N^+-emitter subsystem.

3.5   TEMPERATURE-DEPENDENT PARAMETERS OF NPT IGBT AND THEIR CONNECTION TO THE ELECTRICAL MODEL

During the simulation, the instantaneous power dissipation of a non-punch-through (NPT) IGBT is calculated. The dissipated power is input to the IGBT thermal model, which calculates the IGBT junction temperature. Then the temperature-dependent parameter values of the IGBT are calculated by the equations given in Chapter 1, Section 1.2. A Parameter Temperature IGBT (PTI) subsystem has as its input the IGBT junction temperature, and as its outputs the 11 temperature-dependent material parameter values associated with an IGBT, which are used in the device electrical model. The PTI subsystem block is shown in Fig. 3.22, and its realization in Simulink is given in Fig. 3.23.

A physics-based model of an IGBT, using a Fourier-based solution for the ADE of the lightly doped N^- drift region, including calculation of the temperature-dependent parameter values during every iteration, is presented below. The NPT IGBT shown in Fig. 3.24 has four regions: P emitter (collector terminal), N^- drift, P well, and N^+ emitter. The external (emitter and collector currents, I_E and I_C) and internal electron and hole (at junctions J_1: I_{n_1} and I_{p_1}; and J_2: I_{n_2} and I_{p_2}) and capacitive displacement currents (I_{disp_2} and I_{CG}) are indicated in the figure. An IGBT is a voltage-controlled device. In the normal forward conduction mode of operation, a depletion layer is formed at the MOS-end of the drift region (at the N^- drift region/P well junction), while in reverse-blocking operation, it forms at the P emitter/drift region junction. In forward blocking, a depletion layer is never formed at the P emitter.

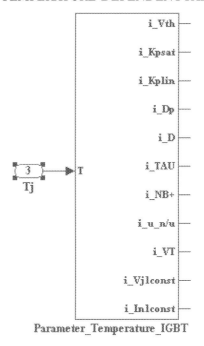

Figure 3.22: A Simulink subsystem for calculating temperature-dependent parameter values for the IGBT.

Figure 3.25 presents the connections of the 11 outputs of the PTI subsystem with the three IGBT subsystems implemented in Simulink.

Similar to what was described for the diode simulation, the temperature-dependent material parameter values are calculated for the IGBT. The calculated temperature-dependent parameters D_p, D, τ, $N_{N^-}^+$, Q_N, and μ are denoted as i_Dp, i_D, i_Tau, i_TAU, i_NB+, and i_u_n/u. As shown in Fig. 3.25, these parameters are inputs to the N-drift region subsystem of the IGBT. Figure 3.26 shows the internal layout of the N^- drift subsystem including the temperature-dependent parameter values implemented in Simulink. It can be seen that the temperature-dependent parameters are fed only to Carrier storage, Feedback, Drift region voltage drop, and Miller capacitance current sub-subsystems. The displacement current sub-subsystem is realized by equations that do not include temperature-dependent parameters.

The Carrier storage sub-subsystem has temperature-dependent parameter inputs of i_D, i_Dp, and i_Tau, as shown in Fig. 3.26. This sub-subsystem provides the solution to the ADE, Eq. (3.15) by using the Fourier solution given by Eq. (3.16) and the boundary conditions of Eq. (3.17). During the forward blocking mode of an IGBT, the depletion layer at junction J_1 will never form. Therefore, its thickness x_1 can be approximated as zero and is constant, so that

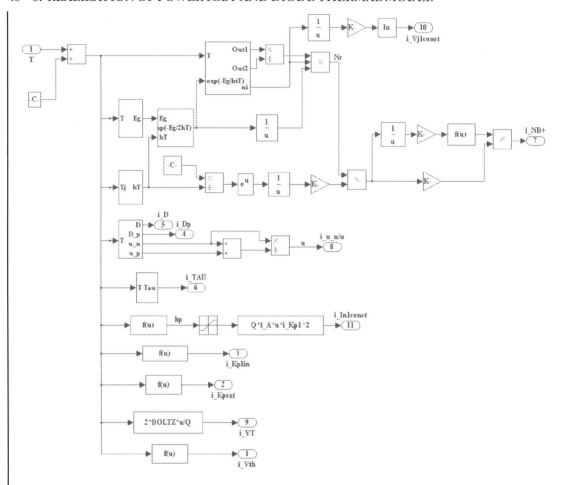

Figure 3.23: Realization of the PTI subsystem in Simulink.

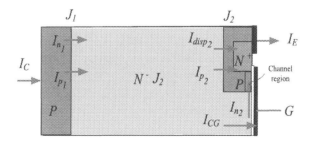

Figure 3.24: One-dimensional cross-section used for modeling the NPT IGBT showing the currents in each region.

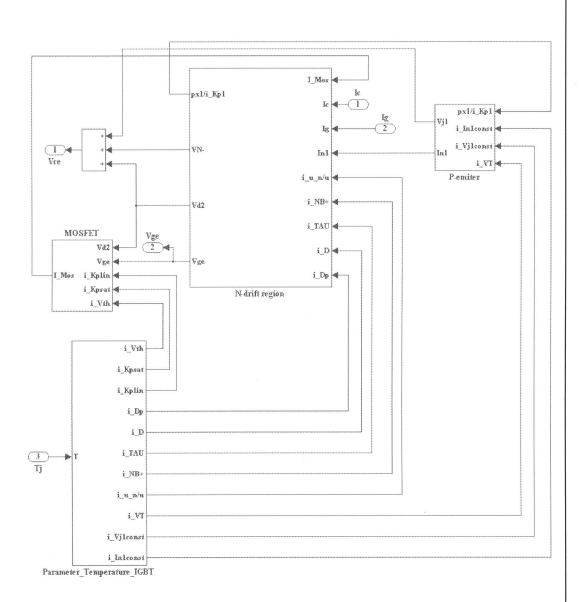

Figure 3.25: PTI subsystem and the three IGBT subsystems in Simulink.

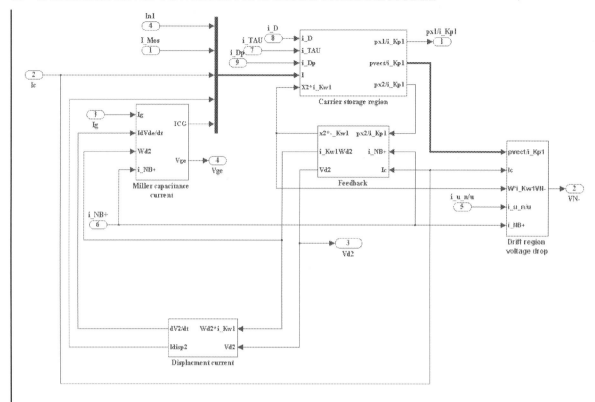

Figure 3.26: The N^- drift region subsystem with temperature-dependent parameters implemented in Simulink.

$dx_1/dt = 0$. Then the Fourier series solution of the ADE given by Eq. (3.16) can be simplified to:

for $k > 0$:

$$D_{N^-}\left[\left.\frac{\partial p(x,t)}{\partial x}\right|_{x2}(-1)^k - \left.\frac{\partial p(x,t)}{\partial x}\right|_{x1}\right] = \frac{x_2}{2}\left(\frac{dp_k(t)}{dt} + \left[\frac{1}{\tau} + \frac{D_{N^-}k^2\pi^2}{x_2{}^2}\right]p_k(t)\right)$$

$$-\left(\sum_{\substack{n=1\\n\neq k}}^{\infty}\frac{n^2}{n^2-k^2}\left[(-1)^{n+k}\frac{dx_2}{dt}\right]p_n(t) + \frac{p_k}{4}\left(\frac{dx_2}{dt}\right)\right) \tag{3.37}$$

for $k = 0$:

$$D_N - \left[\left. \frac{\partial p(x,t)}{\partial x} \right|_{x_2} - \left. \frac{\partial p(x,t)}{\partial x} \right|_{x1} \right] = x_2 \left(\frac{d\, p_0(t)}{dt} + \frac{p_0(t)}{\tau} \right) - \sum_{n=1}^{\infty} \left[(-1)^n \frac{d\, x_2}{dt} \right] p_n(t).$$

Using the current continuity equation at junction J_2:

$$I_C = I_{n_2} + I_{p2} + I_{disp_2} + I_{CG} \tag{3.38}$$

the electron current at junction J_2:

$$I_{n_2} = I_{mos} \tag{3.39}$$

and substituting Eq. (3.39) into (3.38) and then in boundary conditions Eq. (3.17), the left side of the solution of the ADE [Eq. (3.37)], for even and odd numbers of k, can be expressed by the following equations:

$$I_{odd} = -D \left(\left. \frac{\partial p}{\partial x} \right|_{x_2} + \left. \frac{\partial p}{\partial x} \right|_{x_1} \right) = \begin{vmatrix} I_{n_1} \\ I_{mos} \\ I_C \\ I_{disp_2} \\ I_{CG} \end{vmatrix} \frac{1}{2qA} \begin{vmatrix} 2 & -2 & \frac{2D}{D_p} & -\frac{D}{D_p} & -\frac{D}{D_p} \end{vmatrix} \tag{3.40}$$

$$I_{even} = D \left(\left. \frac{\partial p}{\partial x} \right|_{x_2} - \left. \frac{\partial p}{\partial x} \right|_{x_1} \right) = \frac{1}{2qA} \begin{vmatrix} I_{n1} \\ I_{mos} \\ I_C \\ I_{disp_2} \\ I_{CG} \end{vmatrix} \begin{vmatrix} -2 & 2 & 0 & \frac{D}{D_p} & \frac{D}{D_p} \end{vmatrix}. \tag{3.41}$$

When the thermal model is not included as part of the simulation, a Matlab program pre-calculates the constant matrix, i_boudmat2, and the Simulink program uses it for the simulation. When the thermal model is included, since the parameters i_D, i_Dn, and i_Dp are temperature dependent; the matrix, boudmat2, should be calculated during every iteration of the simulation. For example, for six harmonics, the matrix boudmat2 is:

$$boudmat2 = \frac{1}{2qA} \begin{vmatrix} -2 & 2 & 0 & -\frac{D}{D_p} & \frac{D}{D_n} \\ 2 & -2 & -\frac{2D}{D_p} & -\frac{D}{D_p} & -\frac{D}{D_p} \\ -2 & 2 & 0 & -\frac{D}{D_p} & \frac{D}{D_n} \\ 2 & -2 & -\frac{2D}{D_p} & -\frac{D}{D_p} & -\frac{D}{D_p} \\ -2 & 2 & 0 & -\frac{D}{D_p} & \frac{D}{D_n} \\ 2 & -2 & -\frac{2D}{D_p} & -\frac{D}{D_p} & -\frac{D}{D_p} \\ -2 & 2 & 0 & -\frac{D}{D_p} & \frac{D}{D_n} \end{vmatrix}. \tag{3.42}$$

Since the parameters i_D, i_Dn, and i_Dp are temperature dependent, the realization of the matrix i_boudmat2 in Simulink needs to be calculated at every time step during the simulation. In this case, the matrix can be implemented in Simulink as a sum of two matrices multiplied by different constants:

$$\text{boudmat2} = \frac{1}{2qA} \text{i\_Kp1} \left[i\_b1 + \frac{D}{D_p} * i\_b2 \right] \tag{3.43}$$

where i_Kp1 $= 10^{15}$ is a scaling factor, used for converging the simulation and the two matrixes i_b1 and i_b2 are:

$$i\_b1 = \begin{vmatrix} -2 & 2 & 0 & 0 & 0 \\ 2 & -2 & 0 & 0 & 0 \\ -2 & 2 & 0 & 0 & 0 \\ 2 & -2 & 0 & 0 & 0 \\ -2 & 2 & 0 & 0 & 0 \\ 2 & -2 & 0 & 0 & 0 \\ -2 & 2 & 0 & 0 & 0 \end{vmatrix}, \quad i\_b2 = \begin{vmatrix} 0 & 0 & 0 & 1 & 1 \\ 0 & 0 & 2 & -1 & -1 \\ 0 & 0 & 0 & 1 & 1 \\ 0 & 0 & 2 & -1 & -1 \\ 0 & 0 & 0 & 1 & 1 \\ 0 & 0 & 2 & -1 & -1 \\ 0 & 0 & 0 & 1 & 1 \end{vmatrix}. \tag{3.44}$$

The Carrier storage subsystem, including the temperature-dependent parameters, is presented in Fig. 3.27. The main difference between the two carrier storage subsystems, one including and the other not including the temperature-dependent parameters, is that the carrier storage subsystem for the thermal model has an additional subsystem i_boudmat2 that is used to calculate the matrix given by Eq. (3.43).

A sub-subsystem using for calculating the matrix i_boudmat2 is given in Fig. 3.28.

Figure 3.26 shows that the N^- drift region subsystem has another sub-subsystem, Feedback, which also has a temperature-dependent parameter. This parameter is the charge due to the ionized impurity concentration of the lightly doped drift region N_{N-}^+. The implementation of the Feedback subsystem is shown in Fig. 3.29.

The Feedback sub-subsystem includes a Depletion Layer Widths sub-subsystem that uses the ionized impurity concentration of the lightly doped drift region, N_{N-}^+, to calculate the depletion width, W_{d2}, using the Eq. (3.25). The implementation of the Depletion layer widths sub-subsystem is shown in Fig. 3.30.

It can also be noticed, that the Drift region voltage drop sub-subsystem shown in Fig. 3.26 has two temperature-dependent parameters: ionized impurity concentration in the lightly doped drift region, N_{N-}^+(i_NB+), and a mobility ratio calculated by Eq. (3.14), i_u_n/u. The voltage drop in the carrier storage region, V_{N-}, is calculated based on the injected carrier concentration Eq. (3.26). The realization of the Drift region voltage drop sub-subsystem in Simulink is shown in Fig. 3.31.

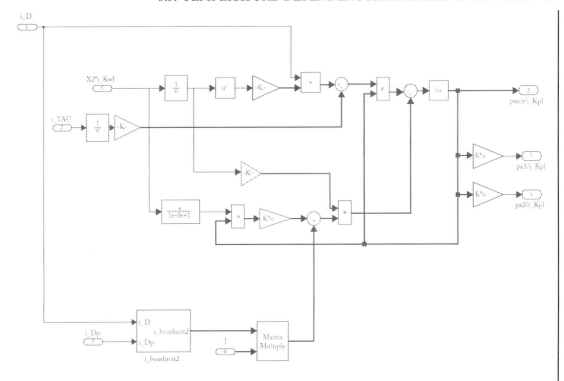

Figure 3.27: Carrier storage subsystem of IGBT.

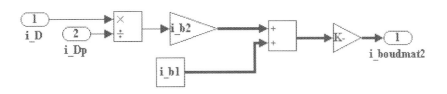

Figure 3.28: Matrix i_boudmat2 realizing in Simulink.

The Miller capacitance current sub-subsystem of the N^- drift region subsystem also includes the temperature-dependent parameter, ionized impurity concentration of the lightly doped drift region, $N_{N^-}^+$.

The Miller capacitance is calculated following [6]. To calculate the Miller capacitance, it is assumed that a depletion layer exists under the gate from P-well and only extends downwards through the drift region once the accumulation layer has been removed from the gate in the center

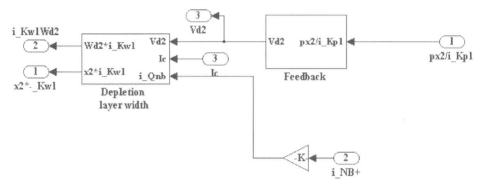

Figure 3.29: Feedback sub-subsystem of IGBT.

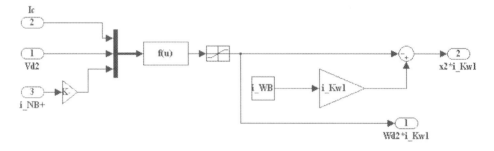

Figure 3.30: Depletion layer widths sub-subsystem.

of the inter-cell region. Therefore, the width of the Miller capacitance, W, is:

$$W = W_{d2} - l_m - l_1 \tag{3.45}$$

where l_m is an inter-cell half-width, and l_1 is a width due to voltage V_{GE}:

$$l_1 = \sqrt{\frac{2\varepsilon V_{GE}}{q N_{N^-}^+}}. \tag{3.46}$$

The Miller capacitance is given by:

$$C_{CG} = \frac{C_{OX} A a_i}{1 + \frac{C_{OX}}{\varepsilon} W} \tag{3.47}$$

where a_i is a ratio of inter-cell area to the total die area.

The current I_{CG} due to Miller capacitance is:

$$I_{CG} = C_{CG} \left(\frac{dV_{d2}}{dt} - \frac{dV_{GE}}{dt} \right) \tag{3.48}$$

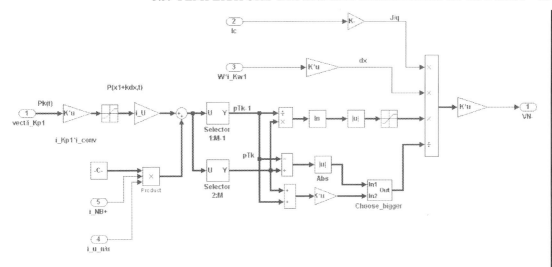

Figure 3.31: The Drift region voltage drop sub-subsystem.

where

$$\frac{dV_{GE}}{dt} = \frac{1}{C_{GE} + C_{GC}} \left(I_G + C_{GC} \frac{dV_{d2}}{dt} \right) . \qquad (3.49)$$

Since the width, l_1, due to voltage V_{GE}, depends on the ionized impurity concentration of the lightly doped drift region, by Eq. (3.46), the Miller capacitance and the current due to it are also dependent on this impurity concentration and, therefore, are temperature dependent. A subsystem calculating the Miller capacitance current is presented in Fig. 3.32.

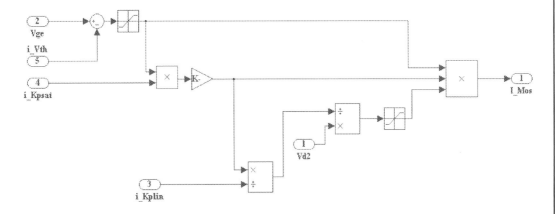

Figure 3.32: Miller capacitance current sub-subsystem.

Another region of the NPT IGBT is the P emitter region. This region is modeled by a subsystem, P-emitter, shown in Fig. 3.25, which is used to calculate the minority current and junction voltage.

Emitter recombination parameter, h_p, is used to calculate the emitter's minority carrier concentration by the following relation:

$$I_{n1} = qAh_p p_{x_1}^2 \qquad (3.50)$$

where p_{x_1} is the excess carrier concentration at junction J_1.

Since the emitter recombination parameter h_p depends on temperature, it too is calculated at each iteration in the PTI subsystem; and the current term i_In1const is used for calculation of the minority current in the P-emitter region. This term is given by the following equation:

$$\text{i\_In1const} = qAh_p \text{i\_Kp1}^2 . \qquad (3.51)$$

The junction voltage across junction J_1 is calculated by the equation:

$$V_{j_1} = 2V_T \ln\left(\frac{p_{x_1}}{n_i}\right) . \qquad (3.52)$$

The two temperature-dependent quantities, i_VT and i_Vj1const, are calculated in the PTI subsystem and used for calculation of the junction voltage V_{J1}. The quantities are calculated by the following equations:

$$\text{i\_VT} = \frac{2kT}{q} , \qquad (3.53)$$

$$\text{i\_Vj1const} = \ln\left(\frac{1}{n_i}\right) . \qquad (3.54)$$

The implementation of the P-emitter subsystem is shown in Fig. 3.33.

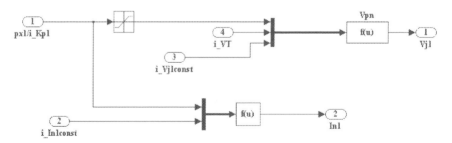

Figure 3.33: P-emitter subsystem.

The final region of the NPT IGBT to be modeled is the MOSFET region associated with the gate contact, gate oxide, and N^+ emitter. It is realized by a subsystem, MOSFET, which is used to calculate the MOSFET current I_{mos}. IGBTs are used as switches, operating mainly in two modes, on and off. During the on-state (conduction), the MOSFET region is operating in the saturation region, while during the off-state (blocking) it operates in cutoff. Therefore, the current, I_{mos}, is calculated by the usual equations describing power MOSFETs:

$$I_{mos} = \begin{cases} \frac{K_p}{2}(V_{GS} - V_{TH})^2 & \text{during on - state} \\ 0 & \text{during off - state} \end{cases} \tag{3.55}$$

where the transconductance coefficient of an IGBT, K_p, is a temperature-dependent parameter, and is also calculated in the PTI subsystem by using Eq. (1.16).

For the simulation of an electro-thermal model of a power semiconductor device, it is recommended that at least two gate pulses be applied (double-pulse test) so the difference in the output parameters, due to self-heating, between the two pulses can be observed. This helps to verify convergence in the simulation and correlation to the correct physical operation of the device/s being simulated.

APPENDIX A

Appendix

As mentioned in Chapter 2, power semiconductor device companies provide a transient thermal impedance response curve to a pulse input of heat power of magnitude 1 kW. An equivalent electrical circuit is provided that has a transient response matched to the measured thermal behavior. The parameters of the circuit are obtained by applying a curve fitting method to the experimental thermal impedance data. The thermal impedance between the semiconductor (junction) and the case of the package can be represented by a finite sum of exponential terms given by the following equation:

$$Z_{thjc} = \sum_{i=1}^{n} R_i \left(1 - e^{-\frac{t}{\tau_i}}\right), \tag{A.1}$$

where i is the term index, R_i and τ_i are correspondingly the thermal resistance and the thermal time constant of the ith term. The thermal time constant, τ_i, can be used for calculating the thermal capacitance, C_i, by following equation:

$$\tau_i = R_i C_i. \tag{A.2}$$

A typical representation will provide values for the thermal resistances and thermal time constants up to order 4 (e.g., i = 1, 2, 3, 4). In this case, a Foster-Equivalent RC thermal network is given in Figure A.1 and has an equivalent driving-point impedance as given in Eqs. (A.1) and (A.2). The network provides only a behavioral description of the system, and does not correlate

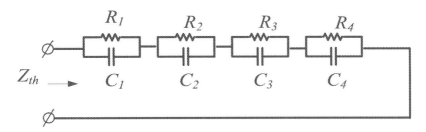

Figure A.1: Foster network of fourth order.

directly with the physical parameters of the package materials and its material geometry. Simulations using a Foster-Equivalent RC thermal network can have difficulty converging. Sometimes it is better to use, instead of the Foster topology, a Cauer network topology that provides the same thermal impedance function and corresponding temperature response. The equivalent

Cauer network can be obtained from the Foster network using the following steps. First, the total driving-point impedance of the fourth-order Foster network is obtained as:

$$Z(s) = \frac{\left(\frac{1}{C_1} + \frac{1}{C_2} + \frac{1}{C_3} + \frac{1}{C_4}\right)s^3 \ldots}{\left(s + \frac{1}{\tau_1}\right)\left(s + \frac{1}{\tau_2}\right)\left(s + \frac{1}{\tau_3}\right)\left(s + \frac{1}{\tau_4}\right)}$$

$$+ \frac{\left[\frac{1}{C_1}\left(\frac{1}{\tau_2} + \frac{1}{\tau_3} + \frac{1}{\tau_4}\right) + \frac{1}{C_2}\left(\frac{1}{\tau_1} + \frac{1}{\tau_3} + \frac{1}{\tau_4}\right) + \frac{1}{C_3}\left(\frac{1}{\tau_1} + \frac{1}{\tau_2} + \frac{1}{\tau_4}\right) + \frac{1}{C_4}\left(\frac{1}{\tau_1} + \frac{1}{\tau_2} + \frac{1}{\tau_3}\right)\right]s^2}{\left(s + \frac{1}{\tau_1}\right)\left(s + \frac{1}{\tau_2}\right)\left(s + \frac{1}{\tau_3}\right)\left(s + \frac{1}{\tau_4}\right)}$$

$$+ \frac{\left[\frac{1}{C_1}\left(\frac{1}{\tau_2\tau_4} + \frac{1}{\tau_3\tau_4} + \frac{1}{\tau_2\tau_3}\right) + \frac{1}{C_2}\left(\frac{1}{\tau_1\tau_3} + \frac{1}{\tau_1\tau_4} + \frac{1}{\tau_3\tau_4}\right) + \frac{1}{C_3}\left(\frac{1}{\tau_1\tau_2} + \frac{1}{\tau_1\tau_4} + \frac{1}{\tau_2\tau_4}\right) + \frac{1}{C_4}\left(\frac{1}{\tau_1\tau_2} + \frac{1}{\tau_1\tau_3} + \frac{1}{\tau_2\tau_3}\right)\right]s \ldots}{\left(s + \frac{1}{\tau_1}\right)\left(s + \frac{1}{\tau_2}\right)\left(s + \frac{1}{\tau_3}\right)\left(s + \frac{1}{\tau_4}\right)}$$

$$+ \frac{\left(\frac{1}{C_1\tau_2\tau_3\tau_4} + \frac{1}{C_2\tau_1\tau_3\tau_4} + \frac{1}{C_3\tau_1\tau_2\tau_4} + \frac{1}{C_4\tau_1\tau_2\tau_3}\right)}{\left(s + \frac{1}{\tau_1}\right)\left(s + \frac{1}{\tau_2}\right)\left(s + \frac{1}{\tau_3}\right)\left(s + \frac{1}{\tau_4}\right)} \qquad (A.3)$$

Then, expanding the denominator into a fourth-order polynomial, the impedance becomes:

$$Z(s) = \frac{\left(\frac{1}{C_1} + \frac{1}{C_2} + \frac{1}{C_3} + \frac{1}{C_4}\right)s^3 \ldots}{divider}$$

$$+ \frac{\left[\frac{1}{C_1}\left(\frac{1}{\tau_2} + \frac{1}{\tau_3} + \frac{1}{\tau_4}\right) + \frac{1}{C_2}\left(\frac{1}{\tau_1} + \frac{1}{\tau_3} + \frac{1}{\tau_4}\right) + \frac{1}{C_3}\left(\frac{1}{\tau_1} + \frac{1}{\tau_2} + \frac{1}{\tau_4}\right) + \frac{1}{C_4}\left(\frac{1}{\tau_1} + \frac{1}{\tau_2} + \frac{1}{\tau_3}\right)\right]s^2}{divider}$$

$$+ \frac{\left[\frac{1}{C_1}\left(\frac{1}{\tau_2\tau_4} + \frac{1}{\tau_3\tau_4} + \frac{1}{\tau_2\tau_3}\right) + \frac{1}{C_2}\left(\frac{1}{\tau_1\tau_3} + \frac{1}{\tau_1\tau_4} + \frac{1}{\tau_3\tau_4}\right) + \frac{1}{C_3}\left(\frac{1}{\tau_1\tau_2} + \frac{1}{\tau_1\tau_4} + \frac{1}{\tau_2\tau_4}\right) + \frac{1}{C_4}\left(\frac{1}{\tau_1\tau_2} + \frac{1}{\tau_1\tau_3} + \frac{1}{\tau_2\tau_3}\right)\right]s}{divider}$$

$$+ \frac{\left(\frac{1}{C_1\tau_2\tau_3\tau_4} + \frac{1}{C_2\tau_1\tau_3\tau_4} + \frac{1}{C_3\tau_1\tau_2\tau_4} + \frac{1}{C_4\tau_1\tau_2\tau_3}\right)}{divider} \qquad (A.4)$$

where *divider* is: $s^4 + \left(\frac{1}{\tau_1} + \frac{1}{\tau_2} + \frac{1}{\tau_3} + \frac{1}{\tau_4}\right)s^3$
$+ \left(\frac{1}{\tau_1\tau_2} + \frac{1}{\tau_1\tau_3} + \frac{1}{\tau_1\tau_4} + \frac{1}{\tau_2\tau_3} + \frac{1}{\tau_2\tau_4} + \frac{1}{\tau_3\tau_4}\right)s^2 + \left(\frac{1}{\tau_1\tau_2\tau_4} + \frac{1}{\tau_1\tau_3\tau_4} + \frac{1}{\tau_1\tau_2\tau_3} + \frac{1}{\tau_2\tau_3\tau_4}\right)s + \frac{1}{\tau_1\tau_2\tau_3\tau_4}$

The impedance can also be expressed in the following compact form:

$$Z(s) = \frac{\alpha_1 s^3 + \alpha_2 s^2 + \alpha_3 s + \alpha_4}{s^4 + \beta_1 s^3 + \beta_2 s^2 + \beta_3 s + \beta_4}, \qquad (A.5)$$

where comparing coefficients from (A.4) and (A.5):

$$\alpha_1 = \left(\frac{1}{C_1} + \frac{1}{C_2} + \frac{1}{C_3} + \frac{1}{C_4} \right) \tag{A.6}$$

$$\alpha_2 = \left[\frac{1}{C_1} \left(\frac{1}{\tau_2} + \frac{1}{\tau_3} + \frac{1}{\tau_4} \right) + \frac{1}{C_2} \left(\frac{1}{\tau_1} + \frac{1}{\tau_3} + \frac{1}{\tau_4} \right) \right. $$
$$\left. + \frac{1}{C_3} \left(\frac{1}{\tau_1} + \frac{1}{\tau_2} + \frac{1}{\tau_4} \right) + \frac{1}{C_4} \left(\frac{1}{\tau_1} + \frac{1}{\tau_2} + \frac{1}{\tau_3} \right) \right] \tag{A.7}$$

$$\alpha_3 = \left[\frac{1}{C_1} \left(\frac{1}{\tau_2 \tau_4} + \frac{1}{\tau_3 \tau_4} + \frac{1}{\tau_2 \tau_3} \right) + \frac{1}{C_2} \left(\frac{1}{\tau_1 \tau_3} + \frac{1}{\tau_1 \tau_4} + \frac{1}{\tau_3 \tau_4} \right) \right. $$
$$\left. + \frac{1}{C_3} \left(\frac{1}{\tau_1 \tau_2} + \frac{1}{\tau_1 \tau_4} + \frac{1}{\tau_2 \tau_4} \right) + \frac{1}{C_4} \left(\frac{1}{\tau_1 \tau_2} + \frac{1}{\tau_1 \tau_3} + \frac{1}{\tau_2 \tau_3} \right) \right] \tag{A.8}$$

$$\alpha_4 = \left(\frac{1}{C_1 \tau_2 \tau_3 \tau_4} + \frac{1}{C_2 \tau_1 \tau_3 \tau_4} + \frac{1}{C_3 \tau_1 \tau_2 \tau_4} + \frac{1}{C_4 \tau_1 \tau_2 \tau_3} \right) \tag{A.9}$$

$$\beta_1 = \left(\frac{1}{\tau_1} + \frac{1}{\tau_2} + \frac{1}{\tau_3} + \frac{1}{\tau_4} \right) \tag{A.10}$$

$$\beta_2 = \left(\frac{1}{\tau_1 \tau_2} + \frac{1}{\tau_1 \tau_3} + \frac{1}{\tau_1 \tau_4} + \frac{1}{\tau_2 \tau_3} + \frac{1}{\tau_2 \tau_4} + \frac{1}{\tau_3 \tau_4} \right) \tag{A.11}$$

$$\beta_3 = \left(\frac{1}{\tau_1 \tau_2 \tau_4} + \frac{1}{\tau_1 \tau_3 \tau_4} + \frac{1}{\tau_1 \tau_2 \tau_3} + \frac{1}{\tau_2 \tau_3 \tau_4} \right) \tag{A.12}$$

$$\beta_4 = \frac{1}{\tau_1 \tau_2 \tau_3 \tau_4} \tag{A.13}$$

From successive steps of synthetic division of: (1) the transfer admittance function, $1/Z(s)$, to determine C_1 of the Cauer network, then (2) the remainder written as an impedance function to determine R_1, etc., we can obtain the full fourth-order ladder network expansion. Figure A.2 shows the resulting Cauer RC network, where the parameters of the circuit are given by the following relationships:

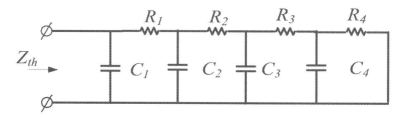

Figure A.2: Cauer RC network of fourth order.

$$C_1 = \frac{1}{\alpha_1} \tag{A.14}$$

$$R_1 = \frac{\alpha_1}{\beta_1 - \frac{\alpha_2}{\alpha_1}} \tag{A.15}$$

$$C_2 = \frac{\gamma_1}{\alpha_1} \tag{A.16}$$

$$R_2 = \frac{\alpha_1 \gamma_2}{\gamma_5} \tag{A.17}$$

$$C_3 = \frac{\gamma_5{}^2}{\alpha_1(\gamma_3 \gamma_5 - \gamma_2 \gamma_6)} \tag{A.18}$$

$$R_3 = \left(\frac{\alpha_1}{\gamma_7}\right)\left(\gamma_3 - \frac{\gamma_2 \gamma_6}{\gamma_5}\right) \tag{A.19}$$

$$C_4 = \frac{\gamma_7}{\alpha_1 \gamma_8} \tag{A.20}$$

$$R_4 = \frac{\alpha_1 \gamma_8}{\beta_4} \tag{A.21}$$

where

$$\gamma_1 = \frac{\left(\beta_1 - \frac{\alpha_2}{\alpha_1}\right)^2}{\left(\beta_1 - \frac{\alpha_2}{\alpha_1}\right)\left(\frac{\alpha_2}{\alpha_1}\right) - \left(\beta_2 - \frac{\alpha_3}{\alpha_1}\right)} \tag{A.22}$$

$$\gamma_2 = \frac{\alpha_2}{\alpha_1} - \frac{\beta_2 - \frac{\alpha_3}{\alpha_1}}{\beta_1 - \frac{\alpha_2}{\alpha_1}} \tag{A.23}$$

$$\gamma_3 = \frac{\alpha_3}{\alpha_1} - \frac{\beta_3 - \frac{\alpha_4}{\alpha_1}}{\beta_1 - \frac{\alpha_2}{\alpha_1}} \tag{A.24}$$

$$\gamma_4 = \frac{\alpha_4}{\alpha_1} - \frac{\beta_4}{\beta_1 - \frac{\alpha_2}{\alpha_1}} \tag{A.25}$$

$$\gamma_5 = \beta_2 - \frac{\alpha_3}{\alpha_1} - \gamma_3 \gamma_1 \tag{A.26}$$

$$\gamma_6 = \beta_3 - \frac{\alpha_4}{\alpha_1} - \gamma_4 \gamma_1 \tag{A.27}$$

$$\gamma_7 = \gamma_6 - \frac{\gamma_5 (\gamma_4 \gamma_5 - \gamma_2 \beta_4)}{\gamma_3 \gamma_5 - \gamma_2 \gamma_6} \tag{A.28}$$

$$\gamma_8 = \gamma_4 - \left(\frac{\gamma_2}{\gamma_5}\right) \beta_4 - \left(\frac{\beta_4}{\gamma_7}\right) \left(\gamma_3 - \frac{\gamma_2 \gamma_6}{\gamma_5}\right) . \tag{A.29}$$

References

[1] A. Caiafa, X. Wang, J.L. Hudgins, E. Santi, and P.R. Palmer, "Cryogenic study and modeling of IGBTs," *IEEE Power Electronics Specialist Conference (PESC) Rec.,* pp. 1897–1903, vol. 4, June 2003, 15–19, Acapulco, Mexico. 2, 3, 4, 5

[2] http://www.efunda.com/materials/materials_home/materials.cfm

[3] G. Slack, R. Tanzilli, R. Pohl, and J. Vandersande, "The Intrinsic Thermal Conductivity of A1N," *J. Phys. Chem. Solids,* vol. 48, no. 7, pp. 641–647, 1987, 472.

[4] B. Du, J.L. Hudgins, E. Santi, A.T. Bryant, P.R. Palmer, and H.A. Mantooth, "Transient electrothermal simulation of power semiconductor devices," *IEEE Trans. Power Electronics,* vol. 5, no. 1, pp. 237–248, January 2010.

[5] https://www.memsnet.org/material/aluminumnitridealnbulk/

[6] P.R. Palmer, E. Santi, J.L. Hudgins, X. Kang, J.C. Joyce, and P.Y. Eng, "Circuit simulator models for the diode and IGBT with full temperature dependent features," *IEEE Trans. Power Electronics,* vol. 18, no. 5, pp. 1220–1229, September 2003. 53

[7] T.K. Gachovska, J.L. Hudgins, Enrico Santi , A. Bryant, and P.R. Palmer, *Modeling Bipolar Power Semiconductor Devices* (Synthesis Lectures on Power Electronics), Morgan & Claypool Publishers; 1st edition (April 5, 2013). 39

Authors' Biographies

TANYA GACHOVSKA

Tanya Kirilova Gachovska received a M.Eng. degree in electrical engineering specializing in Automation of Production and a Ph.D. in electrical engineering specializing in Pulsed Electric Fields (PEF) from the University of Ruse, Bulgaria, in 1995 and 2003, respectively. She worked as an assistant professor from 1999-2003 at the University of Ruse. She conducted research for two years and taught for a semester at McGill University in Montreal, Canada. She had an 18 months post-doc experience on PEF at the University of Nebraska Lincoln, USA. Tanya finished her second Ph.D. in electrical engineering specializing in Power Electronics specifically Modeling of Power Semiconductor Devices in 2012. During her studies, she has taught different courses and labs and continued a collaboration for PEF research with the University of Ruse, McGill, UNL, and two Algerians universities. She is currently at Solantro Semiconductor Inc. She is the author or co-author of more than 30 technical papers and conference presentations, and holds a world patent in PEF.

BIN DU

Bin Du received the Bachelor of Engineer and Master of Engineer degrees in electrical engineering in 2001 and 2004, respectively, from Xi'an Jiaotong University, Xi'an China. In 2008, he received a Ph.D. in electrical engineering from the University of Nebraska-Lincoln. His Ph.D. research focused on electrical and thermal modeling of high-power semiconductor devices.

Since 2008, he has worked with Danfoss Power Electronics, Loves Park, IL as an Electrical Engineer. His job responsibilities include design and qualification on industrial IGBT modules used in high-power, variable frequency drives, IGBT driver design, and paralleling high-power inverter modules.

JERRY L. HUDGINS

Professor Hudgins is a native of West Texas. He attended Texas Tech University, in Lubbock, Texas, where he received a Ph.D. degree in electrical engineering in 1985. Dr. Hudgins served as Associate and Interim Department Chair of Electrical and Computer Engineering at the University of South Carolina prior to joining the University of Nebraska as Chair of the Electrical Engineering Department. Currently, he is Director of the Nebraska Wind Applications Center and Associate Director of the Nebraska Center for Energy Sciences Research. His research involves power electronic device characterization and modeling, power electronics design, and renewable energy systems. In 2000, he was named as an IEEE Third Millenium Medal recipient for "Outstanding Contributions in the area of Power Electronics." Dr. Hudgins is a Fellow of the IEEE and a member of the IEEE Board of Directors (Division II Director). Dr. Hudgins served as the President of the IEEE Power Electronics Society (PELS) for the years of 1997 and 1998 and as President of the IEEE Industry Applications Society (IAS) in 2003. Dr. Hudgins has published over 130 technical papers and book chapters concerning power semiconductors, power electronics, renewable energy systems, and engineering education. He has worked with numerous power semiconductor and equipment manufacturing companies.

ENRICO SANTI

Enrico Santi received a bachelor's degree in electrical engineering from the University of Padua, Padua, Italy, in 1988, and a Ph.D. degree from the California Institute of Technology (Caltech) in 1994. Since 1998, he has been with the Department of Electrical Engineering of the University of South Carolina, where he is currently an Associate Professor. He has published over 100 papers on power electronics and modeling and simulation in international journals and conference proceedings. His research interests include switched-mode power converters, advanced modeling and simulation of power systems, modeling and simulation of semiconductor power devices, and control of power electronic systems. Dr. Santi was the recipient of the National Science Foundation CAREER Award in 2004 and the IEEE Industry Applications Society William M. Portnoy Paper Award in 2003, 2005, and 2006.